# The Missions to Venus
## A Reference Source

# Chapter 1

# The Missions to Venus

## 1.1 Akatsuki (spacecraft)

*Akatsuki* (あかつき, 暁, "Dawn"), also known as the **Venus Climate Orbiter** (**VCO**) and **Planet-C**, is a Japanese (JAXA) space probe tasked to study the atmosphere of Venus. It was launched aboard an H-IIA 202 rocket on 20 May 2010,[*][6] and failed to enter orbit around Venus on 6 December 2010. After the craft orbited the Sun for five years, engineers placed it into an alternative elliptical Venusian orbit on 7 December 2015 by firing its attitude control thrusters for 20 minutes.[*][4][*][5][*][7][*][8] By using five different cameras, *Akatsuki* will study the stratification of the atmosphere, atmospheric dynamics, and cloud physics.[*][9][*][10]

### 1.1.1 Mission

*Akatsuki* is a Japanese space mission to the planet Venus. Planned observations include cloud and surface imaging from an orbit around the planet with an infrared camera, which are aimed at investigation of the complex Venusian meteorology. Other experiments are designed to confirm the presence of lightning and to determine whether volcanism occurs currently on Venus.[*][11] In most planets, the atmosphere circulates much slower than the rotation speed of the planet. However, on Venus, while the planet rotates at 6 km/h at the equator, the atmosphere spins around the planet at 300 km/h.

*Akatsuki* is Japan's first planetary exploration mission since the Nozomi probe, which was launched in 1998 but failed to go into orbit around Mars in 2003 as was planned. *Akatsuki* was originally intended to conduct scientific research for two or more years from an elliptical orbit around Venus ranging from 300 to 80,000 km (190 to 49,710 mi) in altitude,[*][1] but its alternate orbit, yet to be characterized, had to be highly elliptical. The budget for this mission is ¥14.6 billion (US$174 million) for the satellite and ¥9.8 billion (US$116 million) for the launch.[*][12]

### Spacecraft design

The main bus is a 1.45 × 1.04 × 1.44 m (4.8 × 3.4 × 4.7 ft) box with two solar arrays, each with an area of about 1.4 $m^2$ (15 sq ft). The solar arrays provide over 700 W of power in Venus orbit. The total mass of the spacecraft at launch was 517.6 kg (1,141 lb).[*][1] The mass of the science payload is 34 kg (75 lb).[*][13]

Propulsion is provided by a 500-newton (110 $lb_f$) bi-propellant, hydrazine-dinitrogen tetroxide orbital maneuvering engine and twelve mono-propellant hydrazine reaction control thrusters, eight with 23 N (5.2 $lb_f$) of thrust and four with 3 N (0.67 $lb_f$). The total propellant mass at launch was 196.3 kg (433 lb).[*][1]

Communication is via an 8 GHz, 20-watt X-band transponder using the 1.6 m (5 ft 3 in) high-gain antenna. The high-gain antenna is flat to prevent heat from building up in it.[*][10] Akatsuki also has a pair of medium-gain horn antennas mounted on turntables and two low-gain antennas for command uplink. The medium-gain horn antennas are used for housekeeping data downlink when the high-gain antenna is not facing Earth.[*][1]

### Instruments

The scientific payload consists of six instruments: the Lightning and Airglow Camera (LAC), an ultraviolet imager (UVI), a longwave infrared camera (LIR), a 1 μm camera (IR1), a 2 μm camera (IR2), and the radio science (RS) experiment. The five imaging cameras will explore Venus in wavelengths from ultraviolet to the mid-infrared.[*][14]

The LAC will look for lightning in the visible wavelengths of 552 to 777 nanometers. The LIR will study the structure of high-altitude clouds at a wavelength where they emit heat (10 μm). The UVI will study the distribution of specific atmospheric gases such as sulfur dioxide in ultraviolet wavelengths (293–365 nm). The IR1 will peer through semi-transparent windows in Venus' atmosphere to see heat radiation emitted from Venus' surface rocks (0.90–

1.01 μm) and will help researchers to spot active volcanoes, if they exist. The IR2 will detect heat radiation emitted from the lower reaches of the atmosphere (1.65–2.32 μm).[14][15]

### 1.1.2 Public relations

A public relations campaign was held between October 2009 and January 2010 by The Planetary Society and JAXA, to allow individuals to send their name and a message aboard *Akatsuki*.[16][17] Names and messages were printed in fine letters on an aluminum plate and placed aboard *Akatsuki*.[16] 260,214 people submitted names and messages for the mission.[18] Around 90 aluminum plates were created for the spacecraft,[19] including three aluminum plates in which the images of the Vocaloid Hatsune Miku and her super deformed figure Hachune Miku were printed.[20]

### 1.1.3 Operations

**Launch**

*Akatsuki* left the Sagamihara Campus on 17 March 2010, and arrived at the Tanegashima Space Center's Spacecraft Test and Assembly Building 2 on 19 March. On 4 May, *Akatsuki* was encapsulated inside the large payload fairing of the H-IIA rocket that launched the spacecraft, along with the IKAROS solar sail, on a 6-month journey to Venus. On 9 May, the payload fairing was transported to the Tanegashima Space Center's Vehicle Assembly Building, where the fairing was mated to the H-IIA launch vehicle itself.[21]

The spacecraft was launched on 20 May 2010 at 21:58:22 (UTC) from the Tanegashima Space Center,[11] after being delayed because of weather from its initial 18 May scheduled target.[22]

**Orbit insertion failure**

Akatsuki was planned to initiate orbit insertion operations by igniting the orbital maneuvering engine at 23:49:00 on 6 December 2010 UTC.[21] The burn was supposed to continue for twelve minutes, to an initial orbit of 180,000 to 200,000 km (110,000 to 120,000 mi) apoapsis / 550 km (340 mi) periapsis / four days orbital period around Venus.[23]

The orbit insertion maneuver was confirmed to have started on time, but after the expected blackout due to occultation by Venus, the communication with the probe did not recover as planned. The probe was found to be in safe-

*The launch of Akatsuki*

hold mode, spin-stabilized state with ten minutes per rotation.[24] Due to the low communication speed through the low-gain antenna, it took a while to determine the state of the probe.[25] JAXA stated on 8 December that the probe's orbital insertion maneuver had failed.[26][27] At a press conference on 10 December, officials reported that Akatsuki's engines fired for less than three minutes, far short of what was required to enter into Venus orbit.[28] Further research found that the likely reason for the probe malfunction was salt deposits jamming the valve between the helium pressurization tank and the fuel tank. As a result, engine combustion became oxidizer-rich, with resulting high combustion temperatures damaging the combustion chamber throat and nozzle. A similar vapor leakage problem destroyed the Mars Observer probe in 1993.[29]

**Recovery efforts**

JAXA developed plans to attempt another orbital insertion burn when the probe returned to Venus in December 2015. This required placing the probe into "hibernation" or safe mode to prolong its life beyond the original 4.5-year design.

JAXA expressed some confidence in keeping the probe operational, pointing to reduced battery wear, since the probe was then orbiting the Sun instead of its intended Venusian orbit.[30]

Telemetry data from the original failure suggested that the throat of its main engine, the orbit maneuver engine (OME) was still largely intact, and trial jet thrusts of the probe's onboard OME were performed twice, on 7 and 14 September 2011.[21] However, the thrust was only about 40 newtons (9.0 lb$_f$), which was 10% of expectations. Following these tests, it was determined that insufficient specific impulse would be available for orbital maneuvering by the OME. It was concluded that the remaining combustion chamber throat was completely destroyed by transient ignition of the engine. As a result, the selected strategy was to use four hydrazine attitude control thrusters, also called reaction control system (RCS), to drive the probe into orbit around Venus. Because the RCS thrusters do not need oxidiser, the remaining 65 kg of oxidiser (MON) was vented overboard in October 2011 to lighten the spacecraft.[29]

Three peri-Venus orbital maneuvers were executed on 1 November,[11] 10 and 21 November 2011 using the RCS thrusters. A total delta-V of 243.8 m/s was imparted to the spacecraft. Because the RCS thrusters' specific impulse is low compared to the specific impulse of the OME, the previously planned insertion into low Venusian orbit became impossible. Instead, the new plan was to place the probe in a highly elliptical orbit with an apoapsis of a hundred thousand kilometers and a periapsis of a few thousand kilometers from Venus. Engineers planned for the alternate orbit to be prograde (in the direction of the atmospheric superrotation) and lie in the orbital plane of Venus. The method and orbit were announced by JAXA in February 2015, with an orbit insertion date of 7 December 2015.[31] The probe reached its most distant point from Venus on 3 October 2013 and had been approaching the planet since then.[32]

**Orbit insertion**

After performing the last of a series of four trajectory correction manoeuvres between 17 July 2015 and 11 September 2015, the probe was set on the rendezvous trajectory with Venus. The rendezvous occurred on 7 December 2015, when *Akatsuki* was successfully injected into Venusian orbit after a 20-minute burn with four thrusters that were not rated for such a hefty propulsive maneuver.[4][5][33] Instead of taking about 30 hours to complete an orbit around Venus —as was originally planned — *Akatsuki* will complete one orbit every nine days after an adjustment in March 2016.[3]

**Status**

After JAXA engineers measured and calculated its orbit following the 7 December operation, JAXA announced on 9 December that Akatsuki had successfully entered the intended elliptical orbit, as far as 440,000 km (270,000 mi) from Venus, and as close as 400 km (250 mi) from Venus' surface with its orbital period of 13 days and 14 hours.[34] A follow-up thruster burn scheduled for 26 March 2016, will lower *Akatsuki*'s peak altitude of its orbit to about 330,000 km (210,000 mi) and shorten its orbital period from 15 to 9 days.[3] The 2-year science phase will start after the March 2016 orbit adjustment.[3]

Although the spacecraft has survived the orbit transfer, having flown as close as 0.6 AU from the Sun, it is unknown whether the cameras and related electronics aboard have sustained damage, as temperatures within the spacecraft rose 30 °C to 40 °C above design parameters.[29] As a result of high heat flux, the gradual deterioration of heat insulation blankets was noticed, but the deterioration rate slowed in 2015.[35]

### 1.1.4   See also

- Ikaros, solar sail demonstrator, launched along with Akatsuki
- Nozomi (probe), 1998 Mars mission (did not enter orbit)
- Sakigake, Japan's first interplanetary probe, 1985
- Suisei (probe)
- Venus probes

### 1.1.5   References

[1] Takeshi, Oshima; Tokuhito, Sasaki. "Development of the Venus Climate Orbiter PLANET-C (AKATSUKI)". *NEC Technical Journal* **6** (1): 47–51.

[2] Stephen Clark (20 May 2010). "H-2A Launch Report – Mission Status Center". *Spaceflight Now*. Archived from the original on 20 May 2010. Retrieved 20 May 2010.

[3] "Japanese probe fires thrusters in second bid to enter Venus orbit". *The Japan Times*. 7 December 2015. Retrieved 7 December 2015.

[4] Szondy, David. "Akatsuki probe enters orbit around Venus". Retrieved 7 December 2015.

[5] Clark, Stephan. "Japanese probe fires rockets to steer into orbit at Venus". Retrieved 7 December 2015.

[6] Chris Bergin (20 May 2010). "AXA H-IIA carrying Akatsuki and IKAROS launches at second attempt". *NASASpaceFlight*. Retrieved 19 November 2010.

[7] Limaye, Sanjay. "Live from Sagamihara: Akatsuki Orbit Insertion – Second Try". Retrieved 7 December 2015.

[8] Wenz, John (21 September 2015). "Japan's Long Lost Venus Probe May Boom Back to Life". *Popular Mechanics*. Retrieved 14 October 2015.

[9] Nakamura, N.; et al. (May 2011). "Overview of Venus orbiter, Akatsuki". *Earth, Planets and Space* **63** (5): 443–457. doi:10.5047/eps.2011.02.009. ISSN 1880-5981.

[10] "Exploring the Venusian Atmosphere – AKATSUKI/PLANET-C". *Akatsuki Special Site*. Retrieved 5 December 2015.

[11] "AKATSUKI orbit control at perihelion". JAXA. 1 November 2011. Retrieved 3 December 2011.

[12] Staff writers (8 December 2010). "Japan probe shoots past Venus, may meet again in six years". Spacedaily.com. Retrieved 3 December 2011.

[13] "Mission overview". PLANET-C Team/JAXA. Retrieved 3 December 2011.

[14] "Akatsuki (Venus Climate Orbiter / Planet-C)". The Planetary Society. Retrieved 19 November 2010.

[15] Nakamura, Masato; Imamura, Takeshi; Ueno, Munetaka; et al. "Planet-C: Venus Climate Orbiter mission of Japan" (pdf). *Planetary and Space Science* **55** (12): 1831–1842. Bibcode:2007P&SS...55.1831N. doi:10.1016/j.pss.2007.01.009.

[16] "Messages From Earth: Send your Message to Venus on Akatsuki". The Planetary Society. 2010. Archived from the original on 7 April 2010. Retrieved 2 April 2010.

[17] "We will deliver your message to the bright star Venus – Akatsuki Message Campaign". JAXA. Retrieved 19 November 2010.

[18] "AKATSUKI Message Campaign". JAXA. 2010. Retrieved 2 April 2010.

[19] 金星へ届け！県民が寄せ書き[Hoping that It Will Reach Venus! Residents of The Prefecture Write Something Together] (in Japanese). Oita Godo Shimbun. 17 May 2010. Retrieved 20 July 2010.

[20] "打ち上げを目前に控えた「あかつき」と「IKAROS」の機体が公開" [The Airframes of "Akatsuki" And "IKAROS" just before Those Launch Are Opened]. *Mycom Journal* (in Japanese). Mainichi Communications. 12 March 2010. Retrieved 20 July 2010.

[21] "Venus Climate Orbiter "AKATSUKI" (PLANET_C): Topics". JAXA. 1 November 2011. Retrieved 3 December 2011.

[22] "Launch of Venus probe *Akatsuki* postponed due to bad weather". *Japan Today*. 18 May 2010. Retrieved 19 November 2010.

[23] 来月7日に金星周回軌道へ＝あかつき、エンジン噴射 – 7年前は火星で失敗・宇宙機構. *Jiji.com* (in Japanese). Jiji Press. 18 November 2010. Retrieved 5 December 2010.

[24] 金星探査機「あかつき」の状況について[About the State of Venus Probe Akatsuki] (PDF) (in Japanese). 7 December 2010. Retrieved 7 December 2010.

[25] JAXA's press briefing, 22:00, 7 December 2010 JST

[26] "Japan's Venus Probe Fails to Enter Orbit". ABC News. Retrieved 8 December 2010.

[27] "Akatsuki Mission statement". The Planetary Society. Retrieved 8 December 2010.

[28] David Cyranoski (14 December 2010). "Venus miss is a setback for Japanese programme". Nature. Retrieved 21 December 2010.

[29] Nakamura, M.; Kawakatsu, Y.; Hirose, C.; Imamura, T.; Ishii, N.; Abe, T.; Yamazaki, A.; Yamada, M.; Ogohara, K.; Uemizu, K.; Fukuhara, T.; Ohtsuki, S.; Satoh, T.; Suzuki, M.; Ueno, M.; Nakatsuka, J.; Iwagami, N.; Taguchi, M.; Watanabe, S.; Takahashi, Y.; Hashimoto, G. L.; Yamamoto, H. (2014). "Return to Venus of the Japanese Venus Climate Orbiter AKATSUKI". *Acta Astronautica* **93**: 384–389. Bibcode:2014AcAau..93..384N. doi:10.1016/j.actaastro.2013.07.027.

[30] "Japanese Venus Probe Misses Orbit". *Aviation Week & Space Technology*.

[31] "Japanese craft to get second chance after missing Venus in 2010".

[32] 「あかつき」の旅 (2013 年特別公開向け資料) (PDF) (in Japanese). PLANET-C Team/JAXA. 26 August 2013. Retrieved 8 June 2014.

[33] "AKATSUKI: Orbit successfully controlled". PLANET-C Team/JAXA. 5 August 2015. Retrieved 10 September 2015.

[34] "Venus Climate Orbiter "AKATSUKI" Inserted Into Venus' Orbit". JAXA. December 9, 2015.

[35] "AKATSUKI heading to Venus again". PLANET-C Team/JAXA. 9 January 2015. Retrieved 17 March 2015.

## 1.1.6 External links

- JAXA *Akatsuki* Planet-C page
- JAXA *Akatsuki* Special Site
- Akatsuki on Twitter

- Exploring the Venusian Atmosphere – AKATSUKI/PLANET-C – Video

- Launch Report of the H-IIA Launch Vehicle No. 17 with the Venus Climate Orbiter "Akatsuki"(Planet-C) – Video

- Planet-C page (Solar Terrestrial Physics Group)

- Detailed Space Review article about Akatsuki and its recovery

- Presentation about Planet-C from the VEXAG meeting in November 2005 (PDF, 2.7 MB)

- Vieru, Tudor. "JAXA Gets Ready to Launch Venus Probe". Softpedia. Archived from the original on 23 March 2010. Retrieved 30 March 2010.

- Venus Climate Orbiter "AKATSUKI" (PDF, 1.72 Mb)

## 1.2   DAVINCI (spacecraft)

**DAVINCI** (**Deep Atmosphere Venus Investigation of Noble gases, Chemistry, and Imaging**) is a proposed mission concept for an atmospheric probe to Venus.

DAVINCI is planned study the chemical composition of Venus' atmosphere during a 63-minute descent.[*][1][*][2] It will help to determine whether there are active volcanoes on the surface of Venus today and how the surface interacts with the atmosphere of the planet.[*][1]

DAVINCI was selected on 30 September 2015 as one of five semifinalists for Mission #13 of the Discovery Program.[*][1] The winner will be chosen around September 2016,[*][3] and must be ready to launch by the end of 2021.[*][4][*][5]

The principal investigator is Lori Glaze of the Goddard Space Flight Center.

If selected, the mission would be entitled to a 450 million USD budget, and an additional 10 million USD would be awarded if it uses a special 3D woven heat shield.[*][6]

### 1.2.1   Gallery

- Venus as seen in UV light, showing the hazy atmosphere that has complicated traditional surface exploration yet has been the subject of exploration itself. It has produced a number of surprises including the "V" pattern shown here and the high speed winds, intense pressures, and exotic chemistry but the mystery that continues is seen even this image, what is absorbing the UV is unknown even by 2014.[*][1]

1.  ^ [1]

### 1.2.2   See also

- Lucy (spacecraft)

- Near-Earth Object Camera (NEOcam)

- Psyche (spacecraft)

- VERITAS (spacecraft)

### 1.2.3   References

[1] Brown, Dwayne C.; Cantillo, Laurie (30 September 2015). "NASA Selects Investigations for Future Key Planetary Mission". *NASA News* (Washington, D.C.). Retrieved 1 October 2015.

[2] Dreier, Casey; Lakdawalla, Emily (30 September 2015). "NASA announces five Discovery proposals selected for further study". *The Planetary Society*. Retrieved 1 October 2015.

[3] "Small Bodies Dominate NASA's Latest Discovery Competition". *SpaceNews.com*. 7 July 2015. Retrieved 9 August 2015.

[4] Clark, Stephen (24 February 2014). "NASA receives proposals for new planetary science mission". *Space Flight Now*. Retrieved 25 February 2015.

[5] Kane, Van (2 December 2014). "Selecting the Next Creative Idea for Exploring the Solar System". *Planetary Society*. Retrieved 10 February 2015.

[6] Matt Williams, "The DAVINCI spacecraft", *Universe Today*, 6 October 2015.

## 1.3   European Venus Explorer

The **European Venus Explorer (EVE)**, known until 2007 as the **Venus Entry Probe (VEP)**, is a proposed European Space Agency space probe to Venus. In the timeline of the 2005 TRS (technology reference study), the spacecraft was planned to be launched on a Soyuz-2/Fregat launch vehicle around 2013.[*][1] However, submissions to act on the study in 2007[*][2] and 2010[*][3] have been rejected.[*][4]

### 1.3.1   References

[1] ESA description of the *VEP* technology reference study

[2] First proposal for launch in 2016-18 timeframe

[3] Second proposal for 2021-23 launch

[4] EVE - European Venus Explorer

### 1.3.2 External links

- ESA - Venus Entry Probe

- Contractor for the SSTL/ESA – Venus Entry Probe study

- French article mentioning the Venus Entry Probe

## 1.4 High Altitude Venus Operational Concept

*Artist's rendering of a NASA manned floating outpost on Venus*

**High Altitude Venus Operational Concept (HAVOC)** is a NASA study leading to the development of an evolutionary program for the exploration of Venus, with focus on the mission architecture and crewed floating inflatable vehicle concept for missions into Venus's atmosphere.[1][2][3][4]

The HAVOC study comprises a series of missions that would create a Venus colony in five steps:[5]

1. Robotic exploration into the Venusian atmosphere

2. Crewed mission to orbit Venus for 30 days

3. Crewed mission to Venusian atmosphere for 30 days

4. Crewed mission to Venusian atmosphere for 1 year

5. Long-term human presence in a floating balloon.

### 1.4.1 References

[1] "HAVOC". *Systems Analysis and Concepts Directorate of NASA (SACD)*. NASA. Retrieved 1 December 2015.

[2] *A way to explore Venus*. Langley Research Center. 10 October 2015. Retrieved 1 December 2015.

[3] Mack, Eric (20 December 2014). "NASA concept would send astronauts to Venus". *Gizmag*. Retrieved 1 December 2015.

[4] Shadbolt, Peter (3 January 2015). "NASA's plan for an off-world colony: a floating city above Venus". *CNN*. Retrieved 1 December 2015.

[5] Ackerman, Evan (16 December 2014). "NASA Study Proposes Airships, Cloud Cities for Venus Exploration". *IEEE Spectrum*. Retrieved 1 December 2015.

### 1.4.2 External links

- HAVOC Main Project Page

## 1.5 Inspiration Mars Foundation

**Inspiration Mars Foundation** is an American nonprofit organization founded by Dennis Tito that proposed to launch a manned mission to flyby Mars in January 2018,[2][3][4][5][6] or, if the 2018 date is missed, 2021.[7]

The foundation claims that space exploration provides a catalyst for growth, national prosperity, knowledge and global leadership. By taking advantage of this window of opportunity, the Inspiration Mars Foundation intends to revitalize interest in science, technology, engineering and mathematics (STEM) education.

### 1.5.1 History

Some time before the press conference to publicly announce the venture on 27 February 2013, a number of space industry insiders and journalists were given access to some information about the IEEE research paper that would be presented in early March to provide technical details on the feasibility study behind a human crewed free-return mission of 501 days duration in the Mars transfer window of 2018.[3][4][5][6]

On 27 February 2013, the Inspiration Mars Foundation held a press conference in the National Press Club to announce the plan of the foundation to procure space hardware, buy launch vehicle services, select a two-person crew of a married couple[8] (explicitly one man and one woman to represent all of humanity and inspire young persons of both sexes to dream big and pursue science and technology in their schooling), and then attempt to raise the rest of the funding necessary to actually launch a mission in 2018. Dennis Tito is going to fund the foundation on the order of $100 million for its first couple of years of operation.[9]

In comments in November 2013, however, Dennis Tito and others involved with Mars Inspiration indicated that their plan was essentially impossible without significant investment from NASA as well as use of NASA spacecraft.[*][10]

### Funding

The original mission plan was for the mission to be funded entirely through the non-profit foundation that *Inspiration Mars* has in the United States. Dennis Tito planned to fund the foundation's cost for the first "two years from his own deep pockets"[*][11] The total cost of the mission was projected to be between US$1 and US$2 billion,[*][12] less than the US$2.5 billion that NASA is spending on the Mars Science Laboratory robotic rover mission to Mars, including the two years of surface operations via Earth-control of the Curiosity rover. The foundation planned to "raise funds from industry, individuals and others willing to make philanthropic donations".[*][11]

However, in testimony before congress in November 2013, Dennis Tito indicated that he expected private donors to be only able to provide around $300 million in funding, leaving a requirement for an additional investment of $700 million from the US government if the mission was to be feasible. In response NASA stated that whilst they were willing to share technical and programmatic expertise with Inspiration Mars, they are unable to commit to sharing expenses with them.[*][13]

Tito has also indicated that he is unwilling to solicit donations until the mission is in place, saying that "We can't raise money from other donors, and I wouldn't even crowdfund – even from small donors —until we can legitimately say there is a mission on the books . . . And there isn't a mission on the books. We're trying to make that happen."[*][14]

### 1.5.2   Mission

The planned mission[*][9] is a 501-day free-return mission which would allow the spacecraft to use the smallest possible amount of fuel to get it to Mars and back to Earth. "If anything goes wrong, the spacecraft should make its own way back to Earth—but with no possibility of any shortcuts home."[*][15]

In 2018, the planets will align, offering a rare orbit opportunity to travel to Mars and back to Earth in only 501 days. Inspiration Mars intends to send a two-person American crew —a man and a woman—on a journey to within 100 miles of Mars and return them to Earth safely.[*][11]

The mission's first target launch date is 5 January 2018. This quick, free-return orbit opportunity occurs twice every

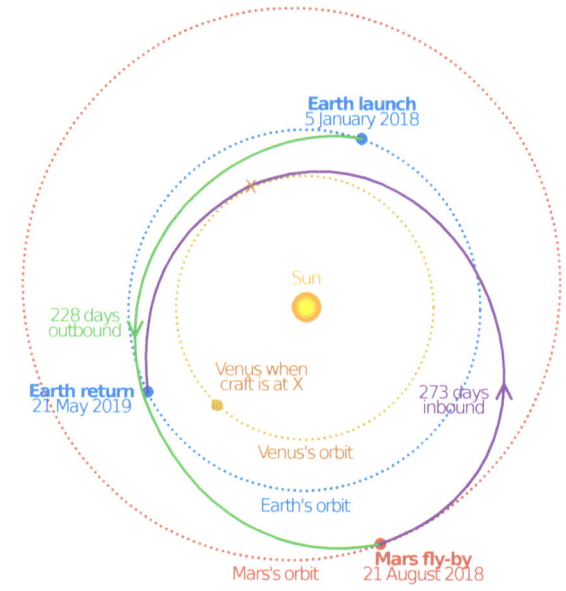

*Approximate Inspiration Mars Trajectory (not to scale)*

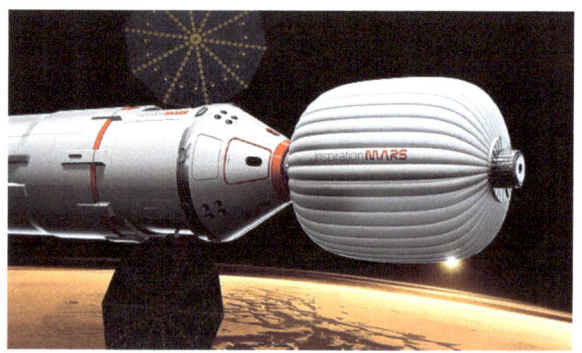

*Artist's Concept of Inspiration Mars Capsule and Hab.*

*Inspiration Mars Periapsis.*

15 years. After 2018, the next opportunity for such a direct trip will not occur again until 2031. By using state-of-the-art technologies derived from NASA and the International Space Station, Inspiration Mars intends to use this oppor-

tunity as a unique platform for science, engineering and STEM education. The science objectives of the mission focus on human endurance and psychology where the mission would set new precedents in human space exploration.[*][15]

An alternate plan, called "Plan B" by Tito, involves a mission that would begin in 2021 but be 88 days longer in duration. It would require both a fly-by of Venus and a fly-by of Mars. This flight would take the craft to within 800 kilometres of the surface of Venus, using the planet in a Gravity assist to speed the onward travel to Mars.[*][16] In comments before congress, Tito described this plan as a "unique trajectory", that would "gives us more time to build the system, and would pass by two planets, Mars and Venus, rather than one" .[*][16] However, Tito also noted that by 2021 other countries may have over-taken the US in the race to get to Mars first.[*][7]

The flyby architecture of either plan lowers risk, with few critical propulsive maneuvers, no entry into the Mars atmosphere, and no rendezvous and docking near Mars. The 2018 plan also represents the shortest duration round-trip mission to Mars. The 2018 launch opportunity coincides with the 11-year solar minimum providing the lowest solar radiation exposure. The next launch opportunity for a direct mission (2031) will not have the advantage of being at the solar minimum, and the 2021 "Plan B" mission would also miss the solar minimum.

When the spacecraft returns to Earth it will enter the atmosphere at 50 000 km per hour (ca 13.9 km/s), faster than any previous return.[*][17] In its initial (or only) contact with the atmosphere it must decrease its speed to less than the escape velocity of Earth, 11.2 km/s.

**Mission technical details**

According to a peer-reviewed paper prepared by Dennis Tito and a group of coauthors for the Institute of Electrical and Electronics Engineers (IEEE), "the mission would require no maneuvers except small course corrections after a trans-Martian injection burn, [and] would allow no aborts. ... [It will] use low-Earth-orbit launch and human-spacecraft technology, outfitted for the long duration of a flight to Mars. The 10-ton crew vehicle—a capsule to best handle the reentry heat and an inflatable or rigid habitat—would contain all of the [life support system or ECLSS] and other gear the crew would need to stay alive. That would include 3,000 pounds (1,400 kg) of dehydrated food, exercise equipment to mitigate the effects of long-term weightlessness, and compact equipment derived from International Space Station gear to recycle water and maintain the atmosphere. There would be no spacesuits or airlock, and the crew would have to endure the travel in about 600 cubic feet (17 $m^3$) of volume." [*][11]

As part of the safety requirements of the mission, and the lack of any space-craft capable of taking everything into orbit in one go,[*][16] the cargo vessel and crew would be launched into Earth orbit separately, and rendezvous in orbit before continuing on to Mars.[*][18] According to Inspiration Mars' chief technology officer, Taber Mcallum, since no commercial rocket is capable of lifting the required mass into orbit in only two launches, this means that use of NASA's Space Launch System is required. However, the SLS is unlikely to be ready for the 2018 launch date.[*][16]

In March 2014, SpaceX indicated that they had been contacted by Inspiration Mars, and were in discussions, but that accommodating such requirements would require some additional work and that such work was not a part of the current focus of SpaceX.[*][19]

**Student design competition**

During the 16th Annual International Mars Society Convention, the Mars Society announced the launch of an international engineering competition for student teams to propose design concepts for the architecture of the Inspiration Mars mission. The contest was open to university engineering student teams from anywhere in the world. "Inspiration Mars is looking for the most creative ideas from engineers all over the world," according to Tito. "Furthermore, we want to engage the explorers of tomorrow with a real and exciting mission, and demonstrate what a powerful force space exploration can be in inspiring young people to develop their talent. This contest will accomplish both of those objectives." [*][20] The design contest took place on August 9th, 2014, and was won by an international team from Purdue and Keio University.[*][21]

**Crew selection**

The foundation is expecting a large number of applications to be the crew for the mission. The mission would likely be record-setting in terms of traveling farther into space than any human has before, and remain in space longer than anyone before. The married couple who is selected will need to "be resilient, even-keel, and able to maintain a happy attitude in the face of adversity", as well as face some health challenges. The year and a half of microgravity will weaken their bodies, and there will be a strong dose of radiation which is not expected to add more than three percent additional risk for fatal cancer, a risk individuals would have to voluntarily accept.[*][8]

As of April 2013, hundreds of "couples who have qualifications that would put them in the running" have offered their services for the mission.[*][22] Much of the initial development work in the early months of the project will be "going to experts in space medicine, life support and thermal protection systems as the team defines the mission. The process includes devising medical, crew-selection and crew-training protocols." The formal call for crew applicants will go out no earlier than 2014.*[22]

### 1.5.3 Challenges

A spokesman for NASA has stated that "Inspiration Mars' s proposed schedule is a significant challenge due to life support systems, space radiation response, habitats and the human psychology of being in a small spacecraft for over 500 days", but that "we remain open to further collaboration as their proposal and plans for a later mission develop" .*[23] John Logsdon, professor emeritus at George Washington University' s Space Policy Institute, has criticised the short time-frame for preparation of the mission, saying that it is "totally implausible" for a mission to be launched in 2018, although the later "Plan B" mission might be possible "if the stars align" .*[18]

The January 2018 launch window is especially likely to be missed because the first launch of the SLS has been delayed by NASA until a likely date of November 2018 or later due to budget concerns.*[24]

### 1.5.4 Foundation management team

- Jonathan Clark, chief medical officer*[25]

- Taber MacCallum, chief technical officer*[25]

- Jane Poynter, developer of the crew and life-support systems*[25]

- Joe Rothenberg, chairman of the advisory and review boards*[25]

- John Carrico, Jr., flight dynamics and trajectory design.*[25]

### 1.5.5 See also

- Deep Space Industries

- Effect of spaceflight on the human body

- Health threat from cosmic rays

- Human spaceflight

- List of 20th-century manned Mars mission plans

- Manned mission to Mars

- Mars Direct

- Mars Settlement

- Mars One

- Mars to Stay

- Private spaceflight

- *The Case for Mars*

### 1.5.6 References

[1] Inspiration Mars Wants To Use ISS, NASAwatch, 15 April 2013

[2] Borenstein, Seth (27 February 2013). "Tycoon wants to send married couple on Mars flyby" . Excite. Associated Press. Retrieved 3 March 2013.

[3] Boucher, Marc (20 February 2013). "The First Human Mission to Mars in 2018 (Updated)". *SpaceRef*. Retrieved 28 February 2013.

[4] Boyle, Alan. "How a millionaire spaceflier intends to send astronauts past Mars in 2018" . *Cosmic Log* (NBC-News.com). Retrieved 28 February 2013.

[5] Mann, Adam (20 February 2013). "Space Tourist to Announce Daring Manned Mars Voyage for 2018" . *Wired*. Retrieved 28 February 2013.

[6] Sonnenberg, Max (23 February 2013). "Millionaire space tourist planning 'historic journey' to Mars in 2018" . *The Space Reporter*. Retrieved 28 February 2013.

[7] "Dennis Tito' s Prepared Remarks Before Congress on Human Mars Mission at Parabolic Arc" . Parabolicarc.com. 2013-11-20. Retrieved 2013-12-07.

[8] Moskowitz, Clara (28 February 2013). "Private Mission to Mars in 2018: Who Should Go?". *space.com*. Retrieved 2 March 2013.

[9] Belfiore, Michael (27 February 2013). "The Crazy Plan to Fly Two Humans to Mars in 2018" . *Popular Mechanics*. Retrieved 28 February 2013.

[10] "Ambitious Mars joy-ride cannot succeed without NASA - space - 21 November 2013" . New Scientist. 2013-11-21. Retrieved 2013-12-07.

[11] Morring, Frank, Jr. (2013-03-04). "Serious Intent About 2018 Human Mars Mission" . *Aviation Week and Space Technology*. Retrieved 2013-03-07.

[12] Koebler, Jason (2013-03-01). "Expert: Dennis Tito's Mars Flyby Has '1-in-3' Chance of Succeeding" . *US News*. Retrieved 2013-03-07. *At a news conference in Washington, D.C., Tito said he's tired of waiting for NASA to send humans to Mars, and that he'd help finance the between $1 and $2 billion needed to complete the mission.*

[13] "Dennis Tito: It will take "less than $1 billion" to make Mars mission happen". Salon.com. 2013-11-20. Retrieved 2013-12-07.

[14] "Clock Ticking for 2018 Private Manned Mars Mission". Space.com. 2013-11-21. Retrieved 2013-12-07.

[15] Connor, Steve (26 February 2013). "The millionaire Dennis Tito and his mission to Mars". *The Independent*. Retrieved 28 February 2013.

[16] "Ambitious Mars joy-ride cannot succeed without NASA - space - 21 November 2013". New Scientist. 2013-11-21. Retrieved 2013-12-07.

[17] Nigel Henbest (13 July 2013). "Race to Mars: Who will be first to the Red Planet?". *New Scientist*: 42–45.

[18] "Going to Mars: Billionaire Dennis Tito plans manned mission with possible 2017 launch". The Washington Post. 2013-06-20. Retrieved 2013-12-07.

[19] Gwynne Shotwell (2014-03-21). *Broadcast 2212: Special Edition, interview with Gwynne Shotwell* (audio file). The Space Show. Event occurs at 11:20–12:10. 2212. Archived from the original (mp3) on 2014-03-22. Retrieved 2014-03-22.

[20] "Rules". The Mars Society. Retrieved 2013-12-07.

[21] http://www.marssociety.org/home/press/announcements/teamkanauwinsinspirationmarsstudentdesigncontest

[22] Morring, Frank (2013-04-12). "Volunteers Line Up For Tito's Mars Flyaround". *Aviation Week*. Retrieved 2013-04-15.

[23] Achenbach, Joel (2011-02-24). "Going to Mars: Billionaire Dennis Tito plans manned mission with possible 2017 launch". The Washington Post. Retrieved 2013-12-07.

[24] Nowakowski, Tomasz. "CHALLENGES FOR ORION AND SLS: AN INTERVIEW WITH GAO DIRECTOR CRISTINA CHAPLAIN". *http://www.spaceflightinsider.com/missions/human-spaceflight/challenges-orion-sls-interview-gao-director-cristina-chaplain/*. Spaceflight Insider. Retrieved 22 December 2014. External link in |website= (help)

[25] Kaufman, Marc (27 February 2013). "Manned Mars Mission Announced by Dennis Tito Group". *National Geographic News*. Retrieved 28 February 2013.

### 1.5.7 External links

- Official website

- Facebook: 'Inspiration Mars' page

- Twitter: @InspirationMars

- Flickr: Inspiration Mars photo stream

- Earth/Mars/IM capsule 501-day mission orbital position animation, Inspiration Mars Foundation, February 2013.

- Feasibility Analysis

- Introductory Press Conference, 27 February 2013

- Inspiration Mars: 29th National Space Symposium: panel discussion, Dennis Tito, Taber MacCallum, Jonathan Clark, and Jane Poynter, Colorado Springs, Colorado, 11 April 2013.

- Newer concept

# 1.6 List of artificial objects on Venus

The following table is a partial **list of artificial objects** on the surface of the planet Venus. They have been abandoned after having served their purpose. The list does not include smaller objects such as parachutes or heatshields.

### 1.6.1 List of artificial objects on Venus

*NASA's Magellan probe radar imagery of Venus surface*

### 1.6.2 Notes

[1] "NASA NSSDC Master Catalog - Venera 3". Retrieved 2010-12-24.

*Venera 9 lander*

### 1.6.3  See also

- List of artificial objects on extra-terrestrial surfaces
- Timeline of planetary exploration

## 1.7  List of missions to Venus

As of 2013, the Soviet Union, United States, European Space Agency and Japan have conducted missions to Venus.

### 1.7.1  References

[1] McDowell, Jonathan. "Launch Log". *Jonathan's Space Page*. Retrieved 21 January 2013.

[2] Krebs, Gunter. "Interplanetary Probes". *Gunter's Space Page*. Retrieved 21 January 2013.

[3] Siddiqi, Asif A. (2002). "1961". *Deep Space Chronicle: A Chronology of Deep Space and Planetary Probes 1958-2000* (PDF). Monographs in Aerospace History, No. 24. NASA History Office. pp. 29–32.

[4] Siddiqi, Asif A. (2002). "1962". *Deep Space Chronicle: A Chronology of Deep Space and Planetary Probes 1958-2000* (PDF). Monographs in Aerospace History, No. 24. NASA History Office. pp. 34–37.

[5] Siddiqi, Asif A. (2002). "1964". *Deep Space Chronicle: A Chronology of Deep Space and Planetary Probes 1958-2000* (PDF). Monographs in Aerospace History, No. 24. NASA History Office. pp. 41–45.

[6] Siddiqi, Asif A. (2002). "1965". *Deep Space Chronicle: A Chronology of Deep Space and Planetary Probes 1958-2000* (PDF). Monographs in Aerospace History, No. 24. NASA History Office. pp. 47–52.

[7] Siddiqi, Asif A. (2002). "1967". *Deep Space Chronicle: A Chronology of Deep Space and Planetary Probes 1958-2000* (PDF). Monographs in Aerospace History, No. 24. NASA History Office. pp. 61–68.

[8] http://sci.esa.int/venus-express/ 55141-venus-express-goes-gently-into-the-night/

### 1.7.2  External links

- Popular Science - February 2003 - Is There Life on Venus? (Google Books link)

## 1.8  Magellan (spacecraft)

The *Magellan* spacecraft, also referred to as the **Venus Radar Mapper**, was a 1,035-kilogram (2,282 lb) robotic space probe launched by NASA on May 4, 1989, to map the surface of Venus by using synthetic aperture radar and to measure the planetary gravitational field.

The *Magellan* probe was the first interplanetary mission to be launched from the Space Shuttle, the first one to use the Inertial Upper Stage booster for launching, and the first spacecraft to test aerobraking as a method for circularizing its orbit. *Magellan* was the fourth successful NASA mission to Venus, and it ended an eleven-year gap in U.S. interplanetary probe launches.

### 1.8.1  History

Beginning in the late 1970s, scientists pushed for a radar mapping mission to Venus. They first sought to construct a spacecraft named the *Venus Orbiting Imaging Radar* (VOIR), but it became clear that the mission would be beyond the budget constraints during the ensuing years. The VOIR mission was canceled in 1982.

A simplified radar mission proposal was recommended by the Solar System Exploration Committee, and this one was submitted and accepted as the Venus Radar Mapper program in 1983. The proposal included a limited focus and a single primary scientific instrument. In 1985, the mission was renamed *Magellan*, in honor of the sixteenth-century Portuguese explorer Ferdinand Magellan, known for his exploration, mapping, and circumnavigation of the Earth.[*][1][*][2][*][3]

The objectives of the mission included:[*][4]

- Obtain near-global radar images of the Venusian surface with a resolution equivalent to optical imaging of 1.0 km per line pair. (*primary*)

- Obtain a near-global topographic map with 50 km spatial and 100 m vertical resolution.

- Obtain near-global gravity field data with 700 km resolution and two to three milligals of accuracy.

- Develop an understanding of the geological structure of the planet, including its density distribution and dynamics.

## 1.8.2 Spacecraft design

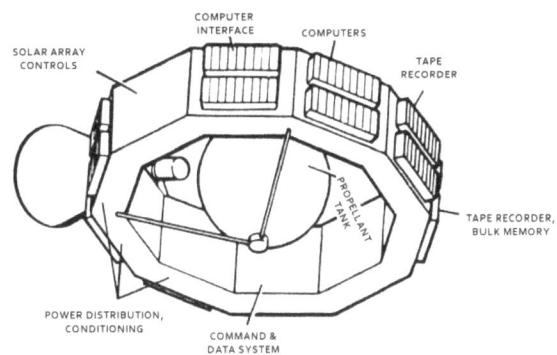

The spacecraft was designed and built by the Martin Marietta Company,[*][5] and the Jet Propulsion Laboratory (JPL) managed the mission for NASA. Elizabeth Beyer served as the program manager and Joseph Boyce served as the lead program scientist for the NASA headquarters. For the JPL, Douglas Griffith served as the *Magellan* project manager and R. Stephen Saunders served as the lead project scientist.[*][1]

To save costs, most of the *Magellan* probe was made up of spare parts from various missions, including the Voyager program, *Galileo*, *Ulysses*, and Mariner 9. The main body of the spacecraft, a spare one from the Voyager missions, was a 10-sided aluminum bus, containing the computers, data recorders, and other subsystems. The spacecraft measured 6.4 meters tall and 4.6 meters in diameter. Overall, the spacecraft weighed 1,035 kilograms and carried 2,414 kilograms of propellant for a total mass of 3,449 kilograms.[*][2][*][6]

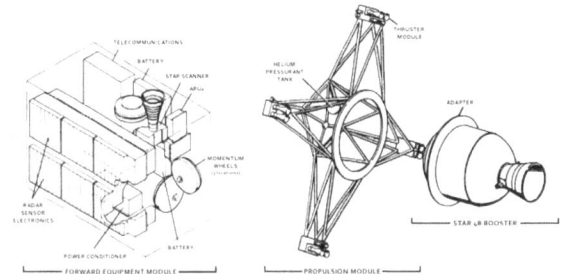

### Attitude control and propulsion

The spacecraft was designed to be three-axis stabilized, including during the firing of the Star 48B solid rocket motor (SRM) used to place it into orbit around Venus. Prior to *Magellan*, all spacecraft SRM firings had involved spinning spacecraft, which made control of the SRM a much easier task. In a typical spin mode, any unwanted forces related to SRM or nozzle mis-alignments are cancelled out. In the case of *Magellan*, the spacecraft design did not lend itself to spinning, so the resulting propulsion system design had to accommodate the challenging control issues with the large Star 48B SRM. The Star 48B, containing 2,014 kg of solid propellant, developed a thrust of ~89,000 Newton (20,000 lbf) shortly after firing; therefore, even a 0.5% SRM alignment error could generate side forces of 445 N (100 lbf). Final conservative estimates of worst-case side forces resulted in the need for eight 445 N thrusters, two in each quadrant, located out on booms at the maximum radius that the Space Shuttle Orbiter Payload Bay would accommodate (4.4-m or 14.5-ft diameter).

The actual propulsion system design consisted of a total of 24 monopropellant hydrazine thrusters fed from a single 71 cm (28 in) diameter titanium tank. The tank contained 133 kg (293 lbm) of purified hydrazine. The design also included a pyrotechnically-isolated external high pressure tank with additional helium that could be connected to the main tank prior to the critical Venus orbit insertion burn to ensure maximum thrust from the 445 N thrusters during the SRM firing. Other hardware regarding orientation of the spacecraft consists of a set of gyroscopes and a star scanner.[*][2][*][3][*][6][*][7]

### Communications

For communications, the spacecraft included a lightweight graphite/aluminum, 3.7-meter high-gain antenna left over from the Voyager Program and a medium-gain antenna spare from the Mariner 9 mission. A low-gain antenna attached to the high-gain antenna, was also included for contingencies. When communicating with the Deep Space

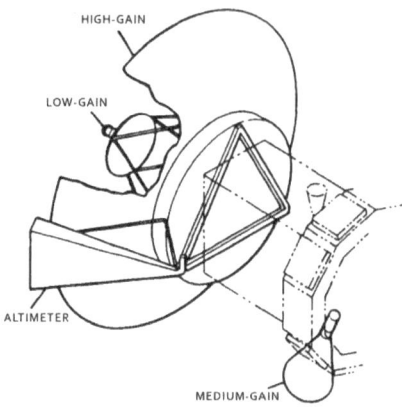

### Scientific instruments

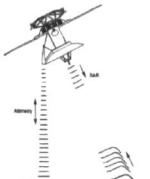

Orientation while collecting data.

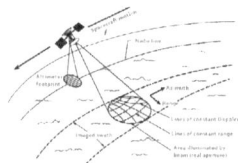

Orbital path for collecting RDRS data.

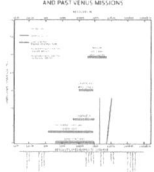

Comparison to previous missions.
The RDRS was a much more capable instrument compared to previous missions.

Network, the spacecraft was able to simultaneously receive commands at 1.2 kilobits/second in the S-band and transmit data at 268.8 kilobits/second in the X-band.[2][3][6][7]

### Power

*Magellan* was powered by two square solar arrays, each measuring 2.5 meters across. Together, the arrays supplied 1,200 watts of power at the beginning of the mission. However, over the course of the mission the solar arrays gradually degraded due to frequent, extreme temperature changes. To power the spacecraft while occulted from the Sun, twin 30 amp-hour, 26-cell, nickel-cadmium batteries were included. The batteries recharged as the spacecraft received direct sunlight.[2][6]

### Computer

The computing system on the spacecraft, derived from the *Galileo* orbiter, included two ATAC-16 computers to control attitude, and four RCA 1802 microprocessors to control the command and data subsystem (CDS). The CDS was able to store commands for up to three days, and also to autonomously control the spacecraft if problems were to arise while mission operators were not in contact with the spacecraft.

For storing the commands and recorded data, the spacecraft also included two multitrack digital tape recorders, able to store up to 225 megabytes of data until contact with the Earth was restored and the tapes were played back.[2][6][7]

Thick and opaque, the atmosphere of Venus required a method beyond optical survey, to map the surface of the planet. The resolution of conventional radar depends entirely on the size of the antenna, which is greatly restricted by costs, physical constraints by launch vehicles and the complexity of maneuvering a large apparatus to provide high resolution data. *Magellan* addressed this problem by using a method known as synthetic aperture, where a large antenna is imitated by processing the information gathered, by ground computers.[8][9]

The *Magellan* high-gain parabolic antenna, oriented 28°–78° to the right or left of nadir, emitted thousands of microwave pulses that passed through the clouds and to the surface of Venus, illuminating a swath of land. The Radar System then recorded the brightness of each pulse as it reflected back off the side surfaces of rocks, cliffs, volcanoes and other geologic features, as a form of backscatter. To increase the imaging resolution, *Magellan* recorded a series of data bursts for a particular location during multiple instances called, "looks". Each "look" slightly overlapped the previous, returning slightly different information for the same location, as the spacecraft moved in orbit. After transmitting the data back to Earth, Doppler modeling was used to take the overlapping "looks" and combine them into a continuous, high resolution image of the surface.[8][9][10]

## 1.8.3 Mission profile

### Launch and trajectory

*Magellan* was launched on May 4, 1989, at 18:46:59 UTC by the National Aeronautics and Space Administration from KSC Launch Complex 39B at the Kennedy Space Center in Florida, aboard Space Shuttle *Atlantis* during mission STS-30. Once in orbit, the *Magellan* and its attached Inertial Upper Stage booster were deployed from *Atlantis* and launched on May 5, 1989 01:06:00 UTC, sending the spacecraft into a Type IV heliocentric orbit where it would circle the Sun 1.5 times, before reaching Venus 15 months later on August 10, 1990.[*][3][*][6][*][7]

Originally, the *Magellan* had been scheduled for launch in 1988 with a trajectory lasting six months. However, due to the Space Shuttle *Challenger* disaster in 1986, several missions, including *Galileo* and *Magellan*, were deferred until shuttle flights resumed in September 1988. *Magellan* was planned to be launched with a liquid-fueled, Centaur-G upper-stage booster, carried in the cargo bay of the Space Shuttle. However, the entire Centaur-G program was canceled after the *Challenger* disaster, and the *Magellan* probe had to be modified to be attached to the less-powerful Inertial Upper Stage. The next best opportunity for launching occurred in October 1989.[*][3][*][6]

Further complicating the launch however, was the launching of the *Galileo* mission to Jupiter, one that included a fly-by of Venus. Intended for launch in 1986, the pressures to ensure a launch for *Galileo* in 1989, mixed with a short launch-window necessitating a mid-October launch, resulted in replanning the *Magellan* mission. Weary of rapid shuttle launches, the decision was made to launch *Magellan* in May, and into an orbit that would require one year, three months, before encountering Venus.[*][3][*][6]

### Orbital encounter of Venus

Main article: Exploration of Venus

Artistic depiction of the orbiter cycle

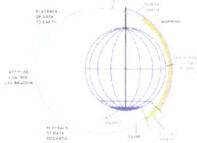

Diagram of the mapping cycle

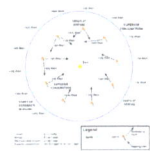

Mapping cycles

The highly elliptical orbit of *Magellan* allowed the high-gain antenna to be used for radar data and communicating with Earth.

On August 7, 1990, *Magellan* encountered Venus and began the orbital insertion maneuver which placed the spacecraft into a three-hour, nine minute, elliptical orbit that brought the spacecraft 295-kilometers from the surface at about 10 degrees North during the periapsis and out to 7762-kilometers during apoapsis.[*][6][*][7]

During each orbit, the space probe captured radar data while the spacecraft was closest to the surface, and then transmit it back to Earth as it moved away from Venus. This maneuver required extensive use of the reaction wheels to rotate the spacecraft as it imaged the surface for 37-minutes and as it pointed toward Earth for two hours. The primary mission intended for the spacecraft to return images of at least 70 percent of the surface during one Venusian day, which lasts 243 Earth days as the planet slowly spins. To avoid overly-redundant data at the highest and lowest latitudes, the *Magellan* probe alternated between a *Northern-swath*, a region designated as 90 degrees north latitude to 54 degrees south latitude, and a *Southern-swath*, designated as 76 degrees north latitude to 68 degrees south latitude. However, due to periapsis being 10 degrees north of the equatorial line, imaging the South Pole region was unlikely.[*][6][*][7]

### Mapping cycle 1

- *Goal: Complete primary objective.*[*][4]

- *September 15, 1990 - May 15, 1991*

The primary mission began on September 15, 1990, with the intention to provide a "left-looking" map of 70% of the Venusian surface at a minimum resolution of 1-kilometer/pixel. During cycle 1, the altitude of the spacecraft varied from 2000-kilometers at the north pole, to 290-kilometers near periapsis. Upon completion during May 15, 1991, having made 1,792 orbits, *Magellan* had mapped approximately 83.7% of the surface with a resolution between 101 to 250-meters/pixel.[*][7][*][15]

### Mission extension

### Mapping cycle 2

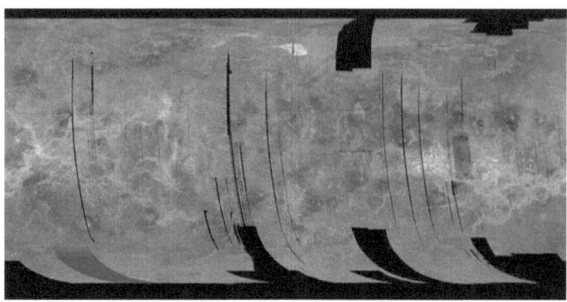

*This map is a mosaic of the "left-looking" data collected during cycle 1.*

- *Goal: Image the south pole region and gaps from Cycle 1.*[16]

- *May 15, 1991 - January 14, 1992*

Beginning immediately after the end of cycle 1, cycle 2 was intended to provide data for the existing gaps in the map collected during first cycle, including a large portion of the southern hemisphere. To do this, *Magellan* had to be reoriented with 180°, changing the gathering method to "right-looking". Upon completion during mid-January 1992, cycle 2 provided data for 54.5% of the surface, and combined with the previous cycle, a map containing 96% of the surface could be constructed.[7][15]

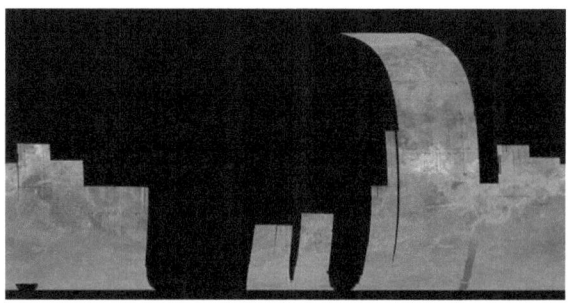

*This map is a mosaic of the "right-looking" data collected during cycle 2.*

## Mapping cycle 3

- *Goal: Fill remaining gaps and collect stereo imagery.*[16]

- *January 15, 1992 - September 13, 1992*

Immediately after cycle 2, cycle 3 began collecting data for stereo imagery on the surface that would later allow the ground team to construct, clear, three-dimensional renderings of the surface. Approximately 21.3% of the surface was imaged in stereo by the end of the cycle on September 13, 1992, increasing the overall coverage of the surface to 98%.[7][15]

## Mapping cycle 4

- *Goal: Measure Venus' gravitational field.*[16]

- *September 14, 1992 - May 23, 1993*

Upon completing cycle 3, *Magellan* ceased imaging the surface. Instead, beginning mid-September 1992, the *Magellan* maintained pointing of the high-gain antenna toward Earth where the Deep Space Network began recording a constant stream of telemetry. This constant signal allowed the DSN to collect information on the gravitational field of Venus by monitoring the velocity of the spacecraft. Areas of higher gravitation would slightly increase the velocity of the spacecraft, registering as a Doppler shift in the signal. The space craft completed 1,878 orbits until completion of the cycle on May 23, 1993; a loss of data at the beginning of the cycle necessitated an additional 10 days of gravitational study.[7][15]

## Mapping cycle 5

- *Goal: Aerobraking to circular orbit and global gravity measurements.*[16]

- *May 24, 1993 - August 29, 1994*

At the end of the fourth cycle in May 1993, the orbit of *Magellan* was circularized using a technique known as aerobraking. The circularized orbit allowed a much higher resolution of gravimetric data to be acquired when cycle 5 began on August 3, 1993. The spacecraft performed 2,855 orbits and provided high-resolution gravimetric data for 94% of the planet, before the end of the cycle on August 29, 1994.[2][3][7][15]

## Aerobraking

- *Goal: To enter a circular orbit*[16]

- *May 24, 1993 - August 2, 1993*

Aerobraking had long been sought as a method for slowing the orbit of interplanetary spacecraft. Previous suggestions included the need for aeroshells that proved too complicated and expensive for most missions. Testing a new approach to the method, a plan was devised to drop the orbit of *Magellan* into the outermost region of the Venusian atmosphere. Slight friction on the spacecraft slowed the velocity over a period, slightly longer than two months, bringing the spacecraft into an approximately circular orbit from

180-kilometers at periapsis to 540-kilometers at apoapsis. The method has since been used extensively on later inter-planetary missions.[*][7][*][15]

**Mapping cycle 6**

- *Goal: Collect high-resolution gravity data and conduct radio science experiments.*[*][16]

- *April 16, 1994 - October 13, 1994*

The sixth and final orbiting cycle was another extension to the two previous gravimetric studies. Toward the end of the cycle, a final experiment was conducted, known as the "Windmill" experiment to provide data on the composition of the upper atmosphere of Venus. *Magellan* performed 1,783 orbits before the end of the cycle on October 13, 1994, when the spacecraft entered the atmosphere and dis-integrated.[*][7]

**Windmill experiment**

- *Goal: Collect data on atmospheric dynamics.*[*][17]

- *September 6, 1994 - September 14, 1994*

In September 1994, the orbit of *Magellan* was lowered to begin the "Windmill experiment". During the experiment, the spacecraft was oriented with the solar arrays broadly, perpendicular to the orbital path, where they could act as paddles as they impacted molecules of the upper-Venusian atmosphere. Countering this force, the thrusters fired to keep the spacecraft from spinning. This provided data on the basic oxygen gas-surface interaction. This was useful for understanding the impact of upper-atmospheric forces which aided in designing future Earth-orbiting satellites, and methods for aerobraking during future planetary space-craft missions.[*][15][*][17][*][18]

**Results**

- Study of the *Magellan* high-resolution global images is providing evidence to better understand Venusian geology and the role of impacts, volcanism, and tectonics in the formation of Venusian surface structures.

- The surface of Venus is mostly covered by volcanic materials. Volcanic surface features, such as vast lava plains, fields of small lava domes, and large shield volcanoes are common.

- There are few impact craters on Venus, suggesting that the surface is, in general, geologically young - less than 800 million years old.

*Rendered image of Venus rotating using data gathered by Magellan.*

*Five global views of Venus by* Magellan.

- The presence of lava channels over 6,000 kilometers long suggests river-like flows of extremely low-viscosity lava that probably erupted at a high rate.

- Large pancake-shaped volcanic domes suggest the presence of a type of lava produced by extensive evolution of crustal rocks.

- The typical signs of terrestrial plate tectonics - continental drift and basin floor spreading - are not evident on Venus. The planet's tectonics is dominated by a system of global rift zones and numerous broad, low domical structures called coronae, produced by the upwelling and subsidence of magma from the mantle.

- Although Venus has a dense atmosphere, the surface reveals no evidence of substantial wind erosion, and only evidence of limited wind transport of dust and sand. This contrasts with Mars, where there is a thin atmosphere, but substantial evidence of wind erosion and transport of dust and sand.

*Magellan* created the first (and currently the best) near-photographic quality, high resolution radar mapping of the planet's surface features. Prior Venus missions had created low resolution radar globes of general, continent-sized formations. *Magellan*, however, finally allowed detailed imaging and analysis of craters, hills, ridges, and other geologic formations, to a degree comparable to the visible-light photographic mapping of other planets. *Magellan*'s global radar map currently remains as the most detailed Venus map in existence, although the planned Russian Venera-D may carry a radar that can achieve the same, if not better resolution as the radar used by *Magellan*.

Media related to Magellan radar imagery at Wikimedia Commons

## 1.8.4   End of mission

*A poster designed for the* Magellan *end of mission.*

On **September 9, 1994**, a press release outlined the termination of the *Magellan* mission. Due to the degradation of the power output from the solar arrays and onboard components, and having completed all objectives successfully, the mission was to end in mid-October. The termination sequence began in late August 1994, with a series of orbital trim maneuvers which lowered the spacecraft into the outermost layers of the Venusian atmosphere to allow the Windmill experiment to begin on September 6, 1994. The experiment lasted for two weeks and was followed by subsequent orbital trim maneuvers, further lowering the altitude of the spacecraft for the final termination phase.[*][17]

On **October 11, 1994**, moving at a velocity of 7 kilometers/second, the final orbital trim maneuver was performed, placing the spacecraft 139.7 kilometers above the surface, well within the atmosphere. At this altitude the spacecraft encountered sufficient ram pressure to raise temperatures on the solar arrays to 126 degrees Celsius.[*][14][*][19]

On **October 13, 1994** at 10:05:00 UTC, communication was lost when the spacecraft entered radio occultation behind Venus. The team continued to listen for another signal from the spacecraft until 18:00:00 UTC, when the mission was determined to have concluded. Although much of *Magellan* was expected to vaporize due to atmospheric stresses, some amount of wreckage is thought have hit the surface by 20:00:00 UTC.[*][14][*][15]

**Quoted from Status Report - October 13, 1994**[*][14]

Communication with the *Magellan* spacecraft was lost early Wednesday morning, following an aggressive series of five Orbit Trim Maneuvers (OTMs) on Tuesday, October 11, which took the orbit down into the upper atmosphere of Venus. The Termination experiment (extension of September "Windmill" experiment) design was expected to result in final loss of the spacecraft due to a negative power margin. This was not a problem since spacecraft power would have been too low to sustain operations in the next few weeks due to continuing solar cell loss.

Thus, a final controlled experiment was designed to maximize mission return. This final, low altitude was necessary to study the effects of a carbon dioxide atmosphere.

The final OTM took the periapsis to 139.7 km (86.8 mi) where the sensible drag on the spacecraft was very evident. The solar panel temperatures rose to 126 deg. C. and the attitude control system fired all available Y-axis thrusters to counteract the torques. However, attitude control was maintained to the end.

The main bus voltage dropped to 24.7 volts

after five orbits, and it was predicted that attitude control would be lost if the power dropped below 24 volts. It was decided to enhance the Windmill experiment by changing the panel angles for the remaining orbits. This was also a preplanned experiment option.

At this point, the spacecraft was expected to survive only two orbits.

*Magellan* continued to maintain communication for three more orbits, even though the power continued to drop below 23 volts and eventually reached 20.4 volts. At this time, one battery went off-line, and the spacecraft was defined as power starved.

Communication was lost at 3:02 AM PDT just as *Magellan* was about to enter an Earth occultation on orbit 15032. Contact was not re-established. Tracking operations were continued to 11:00 AM but no signal was seen, and none was expected. The spacecraft should land on Venus by 1:00 PM PDT Thursday, October 13, 1994.

### 1.8.5 See also

- Venera 15
- Venera 16

### 1.8.6 References

[1] "V-gram. A Newsletter for Persons Interested in the Exploration of Venus" (PDF) (Press release). NASA / JPL. 1986-03-24. Retrieved 2011-02-21.

[2] Guide, C. Young (1990). *Magellan Venus Explorer's Guide*. NASA / JPL. Retrieved 2011-02-22.

[3] Ulivi, Paolo; David M. Harland (2009). *Robotic Exploration of the Solar System Part 2:Hiatus and Renewal 1983–1996*. Springer Praxis Books. pp. 167–195. doi:10.1007/978-0-387-78905-7. Retrieved 2011-02-22.

[4] "Magellan". NASA / National Space Science Data Center. Retrieved 2011-02-21.

[5] Croom, Christopher A.; Tolson, Robert H. "Venusian atmospheric and Magellan properties from attitude control data". *NASA Contractor Report*. NASA Technical Reports Server. Retrieved 8 October 2011.

[6] "SPACE SHUTTLE MISSION STS-30 PRESS KIT" (Press release). NASA. April 1989. Retrieved 2011-02-22.

[7] "Mission Information: MAGELLAN" (Press release). NASA / Planetary Data System. 1994-10-12. Retrieved 2011-02-20.

[8] *Magellan: The unveiling of Venus*. NASA / JPL. 1989. Retrieved 2011-02-23.

[9] Roth, Ladislav E; Stephen D Wall (1995). *The face of Venus : the Magellan radar-mapping mission* (PDF). Washington, D.C.: National Aeronautics and Space Administration. Retrieved 2011-02-21.

[10] Pettengill, Gordon H.; Peter G. Ford; William T. K. Johnson; R. Keith Raney; Laurence A. Soderblom (1991). "Magellan: Radar Performance and Data Products". *Science* (American Association for the Advancement of Science) **252** (5003): 260–5. Bibcode:1991Sci...252..260P. doi:10.1126/science.252.5003.260. JSTOR 2875683. PMID 17769272.

[11] "Synthetic Aperture Radar (SAR)". NASA / National Space Science Data Center. Retrieved 2011-02-24.

[12] "PDS Instrument Profile: Radar System". NASA / Planetary Data System. Retrieved 2011-02-27.

[13] DALLAS, S. S. (1987). "The Venus Radar Mapper Mission". *Acta Astronautica* (Pergamon Journals Ltd) **15** (2): 105–124. doi:10.1016/0094-5765(87)90010-5.

[14] "Magellan Status Report - October 13, 1994" (Press release). NASA / JPL. 1994-10-13. Retrieved 2011-02-22.

[15] Grayzeck, Ed (1997-01-08). "Magellan: Mission Plan". NASA / JPL. Retrieved 2011-02-27.

[16] "Magellan Mission t a Glance" (Press release). NASA. Retrieved 2011-02-21.

[17] "Magellan Begins Termination Activities" (Press release). NASA / JPL. 1994-09-09. Retrieved 2011-02-22.

[18] "Magellan Status Report - September 16, 1994" (Press release). NASA / JPL. 1994-09-16. Retrieved 2011-02-22.

[19] "Magellan Status Report - October 1, 1994" (Press release). NASA / JPL. 1994-10-01. Retrieved 2011-02-22.

### 1.8.7 External links

- Magellan homepage
- *Magellan* mission description and data
- Magellan images
- Magellan Mission Profile by NASA's Solar System Exploration
- http://library.thinkquest.org/J0112188/magellan_probe.htm
- http://nssdc.gsfc.nasa.gov/nmc/spacecraftDisplay.do?id=1989-033B

## 1.9   Manned Venus Flyby

A number of proposals for a **manned Venus flyby** have been considered since the start of the space age.

### 1.9.1   Apollo Applications Program

NASA considered a manned fly-by of Venus in the mid-1960s as part of the Apollo Applications Program, using hardware derived from the Apollo program. Launch would have taken place on October 31, 1973, with a Venus flyby on March 3, 1974 and return to Earth on December 1, 1974.

**Background**

The proposed mission would use a Saturn V to send three men to fly past Venus in a flight which would last approximately one year. The S-IVB stage would be a 'wet workshop' similar to Skylab, first using the S-IVB engine to launch the mission on course to Venus, and then venting of any remaining fuel to serve as home for the crew for the duration of the mission. The Apollo SM engine would be used for course corrections on the way to Venus and back to Earth, and for a braking burn before the Command Module re-entered Earth's atmosphere. In order to free up more space in the Spacecraft Lunar Module Adapter for the docking tunnel connecting the CSM to the S-IVB, the SPS engine on the Service Module would be replaced by two LM engines. These would provide similar thrust with shorter nozzles, and would also give the mission the added safety of redundant engines.

Precursors to the Venus flyby would include an initial orbital test flight with an S-IVB 'wet workshop' and basic docking adapter, and a year-long test flight taking the S-IVB to a near-geostationary orbit around the Earth.

One oddity of the Venus flyby mission is that, unlike trips to the Moon, the CSM would separate and dock with the S-IVB stage before the S-IVB burn, so the astronauts would fly 'eyeballs-out', the thrust of the engine pushing them out of their seats rather than into them. This was required because there was only a short window for an abort burn by the CSM to return to Earth after a failure in the S-IVB, so all spacecraft systems needed to be operational and checked out before leaving the parking orbit around Earth to fly to Venus.

**Science**

The mission would measure:

- Atmospheric density, temperature and pressure as functions of altitude, latitude and time.

- Definition of the planetary surface and its properties.

- Chemical composition of the low atmosphere and the planetary surface.

- Ionospheric data such as radio reflectivity and electron density and properties of cloud layers.

- Optical astronomy - UV and IR measurements above the Earth's atmosphere to aid in the determination of the spatial distribution of hydrogen.

- Solar astronomy - UV, X-ray and possible infrared measurements of the solar spectrum and space monitoring of solar events.

- Radio and radar astronomy - radio observations to map the brightness of the radio sky and to investigate solar, stellar and planetary radio emissions; radar measurements of the surface of Venus and Mercury

- X-ray astronomy - measurements to identify new X-ray sources in the galactic system and to obtain additional information on sources previously identified.

- Data on the Earth-Venus interplanetary environment, including particulate radiation, magnetic fields and meteoroids.

- Data on the planet Mercury, which will be in mutual planetary alignment with Venus approximately two weeks after the Venus flyby

**Mission**

**Phase A**

Phase A of the plan would have launched a 'wet workshop' S-IVB and a standard Block II Apollo CSM into orbit on a Saturn V. The crew would separate the CSM from the S-IVB by blowing off the SLA panels, then perform a Transposition and Docking maneuver similar to that conducted on the lunar flights, in order to dock with the docking module attached to the front of the S-IVB. Optionally they could then use the S-IVB engine to launch them into a high orbit before they vented any remaining fuel into space and entered the S-IVB fuel tanks to conduct experiments for a few weeks. After evaluating the use of the S-IVB as a long-term habitat for astronauts, they would separate the CSM from the S-IVB and return to Earth.

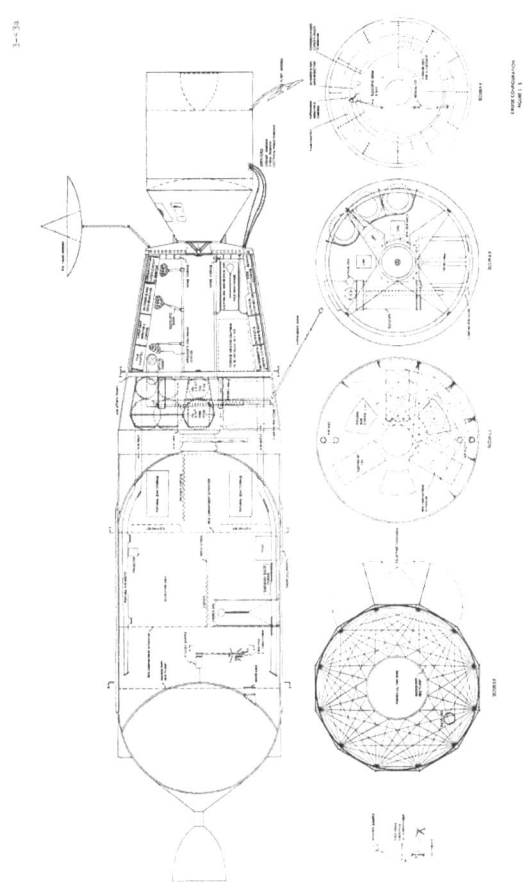

*Cutaway diagram of the Venus flyby spacecraft.*

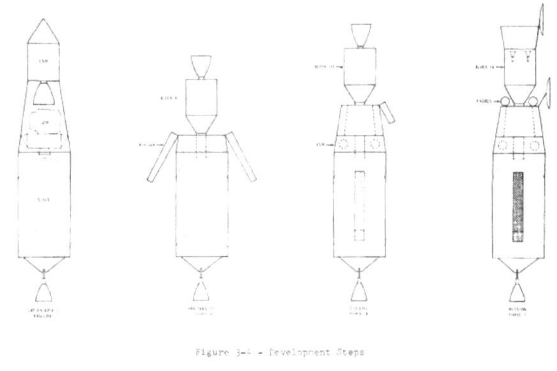

*Diagram of the planned development phases of the Venus flyby spacecraft.*

**Phase B**  Phase B would test the Venus flyby spacecraft in a long duration mission in high orbit. A Saturn V would launch a Block III CSM designed for long-term space-flight and a modified S-IVB with the Environmental Sup-

*Mockup of Phase-A test, in an 'eyeballs-out' burn. (From Orbiter Space Flight Simulator)*

port Module required for the real Venus flyby, and following the transposition and docking maneuver the S-IVB engine would carry the spacecraft to a circular orbit at an altitude of about 25,000 miles around the Earth. This altitude would be high enough to be clear of Earth's radiation belts while exposing the spacecraft to an environment similar to that of a trip to Venus, yet close enough to Earth that the astronauts could use the CSM to return in a few hours in an emergency.

Power would probably be provided by solar panels similar to those used on Skylab, as fuel cells would require a very large amount of fuel to operate for a year. Similarly the fuel cells in the SM used to provide power on lunar flights would be replaced by batteries which would provide enough power for the duration of launch and re-entry operations.

**Phase C**  Phase C would be the actual manned flyby, using a Block IV CSM and an updated version of the Venus flyby S-IVB which would carry a large radio antenna for communication with Earth and two or more small probes which would be released shortly before the flyby to enter the atmosphere of Venus. The Block IV CSM has LM engines replacing the Service Propulsion System engines, batteries to replace the fuel cells, and other modifications to support long-range communication with Earth and the higher re-entry velocities required for the return trajectory compared to a return from lunar orbit.

The Phase C mission was planned to launch in late October or early November 1973, when the velocity requirements required to reach Venus and the duration of the resulting mission would be at their lowest. After a brief stay in Earth parking orbit to check out the spacecraft the crew would head for Venus: in the event of a major problem during the Trans-Venus Injection burn, they would have roughly an hour to separate the CSM from the S-IVB and use the SM engine to cancel out most of the velocity they gained from the burn. This would put them into a highly elliptical orbit which would typically bring them back to Earth for a re-entry two to three days later. Beyond that time the SM engine would not have enough fuel to bring the CSM back to Earth before the SM batteries ran out of power: it would literally be 'Venus or Bust'.

After a successful S-IVB burn, the spacecraft would pass approximately 3000 miles from the surface of Venus about four months later. The flyby velocity would be so high that the crew would only have a few hours for detailed study of the planet. At this point, one or more unmanned probe landers would separate from the main craft and land on Venus.

During the rest of the mission the crew would perform astronomical studies of the Sun and Mercury, which they would approach within 0.3 astronomical units.

### 1.9.2   TMK-MAVR

A variation of the Soviet Union's TMK mission planning for a manned mission to Mars involved a flyby of Venus on the return voyage, and was given the code name "MAVR" (MArs - VeneRa), meaning Mars - Venus. However, the TMK program was cancelled after the N1 rocket that was needed to loft the mission failed to fly successfully.

### 1.9.3   Inspiration Mars

As a "Plan B" in case the proposed Inspiration Mars fly-by of Mars misses its 2018 launch window, in November 2013 Dennis Tito proposed a different mission architecture whereby the mission craft would first fly-by Venus, using Venus in a gravitational slingshot to speed the craft's onward journey to Mars. The proposed mission, which would launch in 2021, would take the Inspiration Mars craft to within 800 kilometres of the Venusian surface.[*][1]

### 1.9.4   See also

- Colonization of Venus

- High Altitude Venus Operational Concept

- Manned mission to Mars

- Orion Asteroid Mission – Proposed Project Constellation mission similar in nature, but to a Near-Earth Asteroid

- TMK (Soviet flyby plan)

### 1.9.5   References

[1] Grossman, Lisa (November 21, 2013). "Ambitious Mars joy-ride cannot succeed without NASA". New Scientist Magazine. Retrieved December 2013.

- *Manned Venus Flyby study*, Feb. 1, 1967

- *Preliminary considerations of Venus exploration via manned flyby, Nov 30, 1967*

- *A Venus lander probe for manned flyby missions*, Feb 23, 1968

- *A survey of manned Mars and Venus flyby missions in the 1970s May 17, 1966*

- *Manned Venus flyby meteorological balloon system, July 29, 1968*

- *Experiment payload for manned venus encounter mission - venus tracking and data orbiter, Jun 13, 1968*

- *Drop sonde and photo sinker probes for a manned venus flyby mission , May 7, 1968*

### 1.9.6   External links

- Human Venus Exploration Architecture Studies

- Manned Venus Orbiting Mission

- NASA Technical Memorandum - Manned Venus Orbiting Mission

- Portree, David S. F. (May 31, 2012). "Piloted Single-Launch Venus Flyby (1967) (Overview of an alternate Manned Venus Flyby Mission)". Beyond Apollo Blog @ Wired.com. Retrieved December 2013.

- Youtube video of the mission, simulated using the Orbiter Spaceflight Simulator

- Manned Venus Flyby: Apollo's Hail Mary Pass, Article at the Failed Steps: The Space Race As It Might Have Been Blog.

## 1.10   Mariner 1

**Mariner 1** was the first spacecraft of the American Mariner program, designed for a planetary flyby of Venus. It was launched aboard a Atlas-Agena rocket on July 22, 1962. Shortly after takeoff the rocket responded improperly to commands from the guidance systems on the ground, setting the stage for an apparent software-related guidance system failure.[*][1] With the craft effectively uncontrolled, a range safety officer ordered its destructive abort 294.5 seconds after launch.[*][2]

According to NASA's current account for the public:

> The booster had performed satisfactorily until an unscheduled yaw-lift (northeast) maneuver was detected by the range safety officer. Faulty application of the guidance commands made steering impossible and were directing the spacecraft towards a crash, possibly in the North Atlantic shipping lanes or in an inhabited area. The destruct command was sent 6 seconds before separation, after which the launch vehicle could not have been destroyed. The radio transponder continued to transmit signals for 64 seconds after the destruct command had been sent.[*][1]

The role of software error in the launch failure remains somewhat mysterious in nature, shrouded in the ambiguities and conflicts among (and in some accounts, even within) the various accounts, official and otherwise. The probe's mission was accomplished by Mariner 2 which launched 5 weeks later.

### 1.10.1 Spacecraft and subsystems

The Mariner 1 spacecraft was identical to Mariner 2, launched 27 August 1962. Mariner 1 consisted of a hexagonal base, 1.04 meters [m] (3.41 ft) across and 0.36 m thick (1.2 ft), which contained six magnesium chassis housing the electronics for the science experiments, communications, data encoding, computing, timing, and attitude control and the power control, battery, and battery charger, as well as the attitude control gas bottles and the rocket engine. On top of the base, was a tall pyramid-shaped mast on which the science experiments were mounted which brought the total height of the spacecraft to 3.66 m (12.0 ft). Attached to either side of the base were rectangular solar panel wings with a total span of 5.05 meters and width of 0.76 meters (16.6 ft × 2.5 ft). Attached by an arm to one side of the base and extending below the spacecraft was a large directional dish antenna.

The Mariner 1 power system consisted of the two solar cell wings, one 183 × 76 cm (72 × 30 in) and the other, 152 × 76 cm (60 × 30 in), with a 31 cm (12 in) dacron extension (a solar sail) to balance the solar pressure on the panels. Those panels powered the craft directly or recharged a 1,000-watt-hour sealed silver-zinc cell battery, which was to be used before the panels were deployed, when the panels were not illuminated by the Sun, and when loads were heavy. A power-switching and booster regulator device controlled the power flow. Communications consisted of

a 3-watt transmitter capable of continuous telemetry operation, the large high gain directional dish antenna, a cylindrical omnidirectional antenna at the top of the instrument mast, and two command antennas, one on the end of either solar panel, which received instructions for midcourse maneuvers and other functions.

Propulsion for midcourse maneuvers was supplied by a monopropellant (anhydrous hydrazine) 225 N retro-rocket. The hydrazine was ignited using nitrogen tetroxide and aluminium oxide pellets, and thrust direction was controlled by four jet vanes situated below the thrust chamber. Attitude control with a 1 degree pointing error was maintained by a system of nitrogen gas jets. The Sun and Earth were used as references for attitude stabilization. Overall timing and control was performed by a digital Central Computer and Sequencer. Thermal control was achieved through the use of passive reflecting and absorbing surfaces, thermal shields, and movable louvers.

The scientific experiments were mounted on the instrument mast and base. A magnetometer was attached to the top of the mast below the omnidirectional antenna. Particle detectors were mounted halfway up the mast, along with the cosmic ray detector. A cosmic-dust detector and solar plasma spectrometer/detector were attached to the top edges of the spacecraft base. A microwave radiometer and an infrared radiometer and the radiometer reference horns were rigidly mounted to a 48 cm (18.9 in) diameter parabolic radiometer antenna mounted near the bottom of the mast.

In addition, a small 91 × 150 cm (3-by-5-foot) United States flag was folded and stowed onboard Mariner 1 (and Mariner 2), before it was mated to the Agena.

### 1.10.2 Launch failure

The launch was aborted due to a combination of two failures, an antenna hardware failure and an onboard guidance system software failure.

First, "the guidance antenna on the Atlas performed poorly, below specifications. When the signal received by the rocket became weak and noisy, the rocket lost its lock on the ground guidance signal that supplied steering commands." [*][3]

As a result, the rocket had to rely on its onboard guidance system, which had a bug in it. There are differing accounts of the details of this error.

**Over bar transcription error**

The most detailed and consistent account was that the error was in hand-transcription of a mathematical symbol in the program specification for the guidance system, in particular a missing overbar.

The error had occurred when a symbol was being transcribed by hand in the specification for the guidance program. The writer missed the superscript bar (or overline) in

$$\bar{R}_h$$

by which was meant "the $n$th smoothed value of the time derivative of a radius R". Since the smoothing function indicated by the bar was left out of the specification for the program, the implementation treated normal minor variations of velocity as if they were serious, causing spurious corrections that sent the rocket off course.*[4]*[5]*[6] It was then destroyed by the Range Safety Officer.*[7]

### Alternate guidance system failure explanations

The cryptic nature of the problems that led to the decision to abort Mariner 1, as well as the confusion in various reports on the incident, led to other explanations in the popular press.

### "The most expensive hyphen in history"

Many accounts note a missing "hyphen" ('-') rather than the over bar, in either the equations, the computer instructions or the data. For example, Arthur C. Clarke wrote several years later that Mariner 1 was "wrecked by the most expensive hyphen in history".*[8]

Several factors contributed to the "missing hyphen" narrative and its longevity, even in official accounts from technical cognoscenti at JPL and NASA. Among the factors cited (or obvious enough):

- The overbar's resemblance to a hyphen ('⁻' versus '-').

- The difficulty of explaining the real error to the American public and its elected representatives.

- External political pressures and internal schedule pressures, as the mission was

  - an expensive failure of a three-way collaboration (JPL, NASA, USAF),

  - legitimized within the narrative of the US-USSR space race,

  - very high profile, as America's first planetary mission,

  - on a very tight schedule, as it was planned with a narrow launch window (45 days), leaving little

time for inquiries, investigations or recriminations before the launch of Mariner 2. The official accounts (which included mentions of a missing hyphen) were the results of an inquiry conducted in less than a week.

Regardless of whatever may have given rise to initial reports of a "missing hyphen", the simplest and most consistent-sounding explanation that the public and Congress would accept would probably have been preferable to those who simply wanted to get on with the job of a Venus fly-by mission. The stories had contradictions, perhaps, but they were so technical that nobody who could have interfered with Mariner-program progress was likely to care about them or even notice. (After all, even in one later NASA account, the supposed "hyphen" is reported as missing from instructions at one point in the text, and from equations at another*[3]).

**Ambiguity of error location**    *The New York Times*, reporting on the results of a review board, said that the error stemmed from "the omission of a hyphen in some mathematical data".*[9] The same report also said the hyphen was

> a symbol that should have been fed into a computer, along with a mass of other coded mathematical instructions.

This sort of inconsistency or ambiguity was seen in many subsequent variations on the story, official and otherwise. "Missing hyphen" versions of the story gained from official support before the month was out. NASA official Richard B. Morrison testified before Congress that the supposed hyphen

> ... gives a cue for the spacecraft to ignore the data the computer feeds it until radar contact is once again restored. When that hyphen is left out, false information is fed into the spacecraft control systems. In this case, the computer fed the rocket in hard left, nose down and the vehicle obeyed and crashed.*[10]

(Note that Morrison says the spacecraft "crashed", not that it was intentionally destroyed). In a NASA account submitted to Congress in 1963, the hyphen is described as missing in two different ways:

> NASA-JPL-USAF Mariner R-1 Post-Flight Review Board determined that the omission of a hyphen *in coded computer instructions* transmitted incorrect guidance signals to Mariner spacecraft

boosted by two-stage Atlas-Agena from Cape Canaveral on July 21. Omission of hyphen *in data editing* caused computer to swing automatically into a series of unnecessary course correction signals which threw spacecraft off course so that it had to be destroyed.*[11]

In the same 1963 report to Congress, Morrison's testimony from the previous year is recounted differently:

> In testimony before House Science and Astronautics Committee, Richard B. Morrison, NASA's Launch Vehicles Director, testified that an error *in computer equations* for Venus probe launch of Mariner R-1 space-craft on July 21 led to its destruction when it veered off course.*[12]

JPL's Mariner Venus Final Project Report in 1965 noted that, at 4 minutes and 25 seconds into the flight, there was an "[U]nscheduled yaw-lift maneuver":

> ...steering commands were being supplied, but *faulty application of the guidance equations* was taking the vehicle far off course.*[13]

In a NASA report published in 1985, Oran Nicks offered another slightly differing account, but with the software-related error still identified as a missing "hyphen":

> The guidance antenna on the Atlas performed poorly, below specifications. When the signal received by the rocket became weak and noisy, the rocket lost its lock on the ground guidance signal that supplied steering commands. The possibility had been foreseen; in the event that radio guidance was lost the internal guidance computer was supposed to reject the spurious signals from the faulty antenna and proceed on its stored program, which would probably have resulted in a successful launch. At this point a second fault took effect. Somehow a hyphen had been dropped from the guidance program loaded aboard the computer, allowing the flawed signals to command the rocket to veer left and nose down. The hyphen had been missing on previous successful flights of the Atlas, but that portion of the equation had not been needed since there was no radio guidance failure.*[3]

NASA's website now says the problem was:

> ... apparently caused by a combination of two factors. Improper operation of the Atlas airborne

beacon equipment resulted in a loss of the rate signal from the vehicle for a prolonged period. The airborne beacon used for obtaining rate data was inoperative for four periods ranging from 1.5 to 61 seconds in duration. Additionally, the Mariner 1 Post Flight Review Board determined that the omission of a hyphen *in coded computer instructions in the data-editing program* allowed transmission of incorrect guidance signals to the spacecraft. During the periods the airborne beacon was inoperative the omission of the hyphen *in the data-editing program* caused the computer to incorrectly accept the sweep frequency of the ground receiver as it sought the vehicle beacon signal and combined this data with the tracking data sent to the remaining guidance computation. This caused the computer to swing automatically into a series of unnecessary course corrections with erroneous steering commands which finally threw the spacecraft off course.*[14]

**Other punctuation**  In other accounts, the bug consisted of:

- a period typed in place of a comma, causing a FORTRAN DO loop statement to be misinterpreted (although there is no evidence that FORTRAN was used in the mission), of the form "DO 5 K=1. 3" interpreted as assignment "DO5K = 1.3"*[15] There are anecdotal reports that there was in fact such a bug in a NASA orbit computation program at about this time, but it was a program for Project Mercury, not Mariner, and the claim was that the bug was noticed and fixed before there could be any serious consequences.*[16]

- a missing comma *[17]

- an extraneous semicolon *[18]

## 1.10.3  References

[1]  "Mariner 1". 4.0.8. NASA. 2008-08-05. Retrieved 2009-02-14.

[2]  "Venus Shot Fails as Rocket Strays" (fee required). New York Times. 1962-07-23. Retrieved 2009-02-14.

[3]  NASA publication SP-480, *Far Travelers -- The Exploring Machines*, Oran W. Nicks, 1985

[4]  Peter Neumann (1989-05-27). "Mariner I -- no holds BARred". *The Risks Digest Volume 8: Issue 75*. Retrieved 2014-10-31.

[5]  . ISBN 978-0262530828. Missing or empty |title= (help)

[6] http://www.faqs.org/faqs/space/probe/

[7] *Beyond the Limits: Flight Enters the Computer Age*, Paul E. Ceruzzi, p.203. In one of the notes for this book (p. 250), the author writes "The same flawed program had been used in earlier Ranger launches with no ill effects."

[8] *The Promise of Space*, Arthur C. Clarke, 1968, p. 225.

[9] "For Want of Hyphen Venus Rocket Is Lost" , *New York Times*, July 27, 1962, as quoted in *RISKS Digest*, Vol 5, Issue #66.

[10] House Science and Astronautics Committee, July 31, 1962, also quoted here

[11] "Astronautical and Aeronautical Events of 1962," report to the House Committee on Science and Astronautics, June 12, 1963 p.131.

[12] "Astronautical and Aeronautical Events of 1962," report to the House Committee on Science and Astronautics, June 12, 1963 p.333

[13] *Mariner Venus Final Project Report* (NASA SP-59, 1965), p.87.

[14] "Mariner 1" , Version 4.0.7, 2 April 2008.

[15] *Beyond the Limits: Flight Enters the Computer Age*, Paul E. Ceruzzi, In p.250, footnote 13 for Chapter 9, where Ceruzzi writes that, "[S]ince the Atlas Guidance Computer did not have a Fortran compiler ...." , and in footnote 14, "The Atlas Launch computer did not even use Fortran, the Fortran programming language. How the story has become embellished in this way is a mystery."

[16] *RISKS Digest*, v. 9, issue 54, "Mariner I [once more]", Mark Brader, 12 December 1989.

[17] Famous bugs

[18] JPL 101 page 22

### 1.10.4   External links

- NASA's article about the Mariner I

- Mariner 1 Mission Profile by NASA's Solar System Exploration

- RISKS Digest detail about the Mariner I failure

# 1.11   Mariner 10

**Mariner 10** was an American robotic space probe launched by NASA on November 3, 1973, to fly by the planets Mercury and Venus.

Mariner 10 was launched approximately two years after Mariner 9 and was the last spacecraft in the Mariner program (Mariner 11 and 12 were allocated to the Voyager program and redesignated Voyager 1 and Voyager 2).

The mission objectives were to measure Mercury's environment, atmosphere, surface, and body characteristics and to make similar investigations of Venus. Secondary objectives were to perform experiments in the interplanetary medium and to obtain experience with a dual-planet gravity assist mission. Mariner 10's science team was led by Bruce C. Murray at the Jet Propulsion Laboratory.[2]

### 1.11.1   Design and trajectory

Mariner 10 was the first spacecraft to make use of an interplanetary gravitational slingshot maneuver, using Venus to bend its flight path and bring its perihelion down to the level of Mercury's orbit.[3] This maneuver, inspired by the orbital mechanics calculations of the Italian scientist Giuseppe Colombo, put the spacecraft into an orbit that repeatedly brought it back to Mercury. Mariner 10 used the solar radiation pressure on its solar panels and its high-gain antenna as a means of attitude control during flight, the first spacecraft to use active solar pressure control.

The components on Mariner 10 can be categorized into four groups based on their common function. The solar panels, power subsystem, attitude control, and computer kept the spacecraft operating properly during the flight. The navigational system, including the hydrazine rocket, would keep Mariner 10 on track to Venus and Mercury. Several scientific instruments would collect data at the two planets. Finally, the antennas would transmit this data to the Deep Space Network back on Earth, as well as receive commands from Mission Control. Mariner 10's various components and scientific instruments were attached to a central hub, which was roughly the shape of an octagonal prism. The hub stored the spacecraft's internal electronics.[1][4][5] The Mariner 10 spacecraft was manufactured by Boeing.[6] NASA set a strict limit of $98 million for Mariner 10's total cost, which marked the first time the agency subjected a mission to an inflexible budget constraint. No overruns would be tolerated, so mission planners carefully considered cost efficiency when designing the spacecraft's instruments.[7] Cost control was primarily accomplished by executing contract work closer to the launch date than was recommended by normal mission schedules, as reducing the length of available work time increased cost efficiency. Despite the rushed schedule, very few deadlines were missed.[8] The mission ended up about $1 million under budget.[9]

Attitude control is needed to keep a spacecraft's instruments and antennas aimed in the correct direction.[10]

During course maneuvers, a spacecraft may need to rotate so that its rocket faces the proper direction before being fired. Mariner 10 determined its attitude using two optical sensors, one pointed at the Sun, and the other at a bright star, usually Canopus; additionally, the probe's three gyroscopes provided a second option for calculating the attitude. Nitrogen gas thrusters were used to adjust Mariner 10's orientation along three axes.[11][12][13] The spacecraft's electronics were intricate and complex: it contained over 32,000 pieces of circuitry, of which resistors, capacitors, diodes, microcircuits, and transistors were the most common devices.[14] Commands for the instruments could be stored on Mariner 10's computer, but were limited to 512 words. The rest had to be broadcast by the Mission Sequence Working Group from Earth.[15] The power subsystem could store up to 20 ampere hours of electricity on a 39 volt nickel-cadmium battery.[16]

The flyby past Mercury posed major technical challenges for scientists to overcome. Due to Mercury's proximity to the Sun, Mariner 10 would have to endure 4.5 times more solar radiation than when it departed Earth—compared to previous Mariner missions, spacecraft parts needed extra shielding against the heat. Thermal blankets and a sunshade were installed on the main body. After evaluating different choices for the sunshade cloth material, mission planners chose beta cloth, a combination of aluminized Kapton and glass-fiber sheets treated with Teflon.[17] However, solar shielding was unfeasible for some of Mariner 10's other components. Mariner 10's two solar panels needed to be kept under 115 °C. Covering the panels would defeat their purpose of producing electricity. The solution was to add an adjustable tilt to the panels, so the angle at which they faced the sun could be changed. Engineers considered folding the panels toward each other, making a V-shape with the main body, but tests found this approach had the potential to overheat the rest of the spacecraft. The alternative chosen was to mount the solar panels in a line and tilt them along that axis, which had the added benefit of increasing the efficiency of the spacecraft's nitrogen jet thrusters, which could now be placed on the panel tips. The panels could be rotated a maximum of 76 degrees.[5][18] Additionally, Mariner's 10 hydrazine rocket nozzle had to face the Sun to function properly, but scientists rejected covering the nozzle with a thermal door as an undependable solution. Instead, a special paint was applied to exposed parts on the rocket so as to reduce heat flow from the nozzle to the delicate instruments on the spacecraft.[19]

Accurately performing the gravity assist at Venus posed another hurdle.[20] If Mariner 10 was to maintain a course to Mercury, its trajectory could deviate no more than 200 kilometres (120 mi) from a critical point in the vicinity of Venus.[21] To ensure that the necessary course corrections could be made, mission planners tripled the amount of hydrazine fuel Mariner 10 would carry, and also equipped the spacecraft with more nitrogen gas for the thrusters than the previous Mariner mission had held. These upgrades proved crucial in enabling the second and third Mercury flybys.[22]

Even so, the mission still lacked the ultimate safeguard: a sister spacecraft. It was common for probes to be launched in pairs, with complete redundancy to guard against the failure of one or the other.[23] The budget constraint ruled this option out. Even though mission planners stayed sufficiently under budget to divert some funding for constructing a backup spacecraft, the budget did not permit both to be launched at the same time. In the event that Mariner 10 failed, NASA would only allow the backup to be launched if the fatal error was diagnosed and fixed—this would have to be completed in the two-and-a-half weeks between the scheduled launch on November 3 and the last possible launch date of November 21.[22][24]

## 1.11.2 Instruments

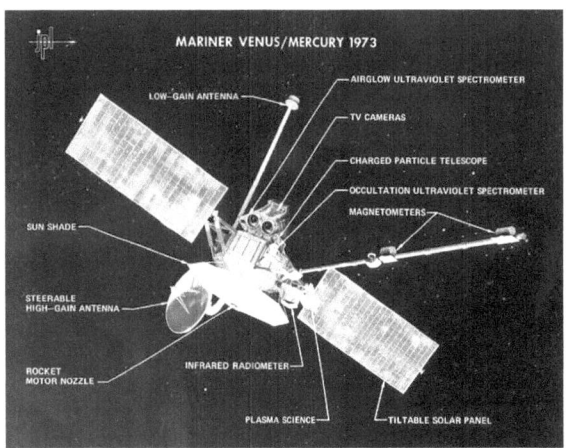

*An illustration showing Mariner 10's instruments*

Mariner 10 conducted seven experiments at Venus and Mercury. Six of these experiments had a dedicated scientific instrument to collect data.[25] The experiments and instruments were designed by research laboratories and educational institutions from across the United States.[26] Out of forty-six submissions, JPL selected seven experiments on the basis of maximizing science return without exceeding cost guidelines: together, the seven scientific experiments cost 12.6 million dollars, about one-eighth of the total mission budget.[8]

**Television photography**

The imaging system, the Television Photography Experiment, consisted of two 15 cm (5.9″) Cassegrain telescopes feeding vidicon tubes.[27] The main telescope could be bypassed to a smaller wide angle optic, but using the same tube.[27] It had an 8-position filter wheel, with one position occupied by a mirror for the wide-angle bypass.[27]

The entire imaging system was imperiled when electric heaters attached to the cameras failed to turn on immediately after launch. To avoid the Sun's damaging heat, the cameras were deliberately placed on the spacecraft side facing away from the Sun. Consequently, the heaters were needed to prevent the extremely cold environment from harming the cameras. JPL engineers found that the vidicons could generate enough heat through normal operation to stay just above the critical temperature of −40 °C; therefore they advised against turning off the cameras during the flight. Fortunately, test photos of the Earth and Moon showed that image quality had not been significantly affected.[28] The mission team was pleasantly surprised when the camera heaters started working on January 17, two months after launch.[29][30] Later investigation concluded that a short circuit in a different location on the probe had prevented the heater from turning on. This allowed the vidicons to be turned off as needed.[31]

Of the six main scientific instruments, the 43.6 kilograms (96 lb) cameras were by far the most massive device. Requiring 67 watts of electricity, the cameras consumed more power than the other five instruments combined.[32] The system returned about 7000 photographs of Mercury and Venus during Mariner 10's flybys.[27]

**Infrared radiometer**

The infrared radiometer detected infrared radiation given off by Mercury's surface and Venus' atmosphere, from which the temperature could be calculated. How quickly the surface lost heat as it rotated into the planet's dark side revealed aspects about the surface's composition, such as whether it was made out of rocks, or out of finer particles.[33][34] The infrared radiometer contained a pair of Cassegrain telescopes fixed at an angle of 120 degrees to each other, and a pair of detectors made from antimony-bismuth thermopiles. The instrument was designed to measure temperatures as cold as −193 °C and as hot as 427 °C. Stillman C. Chase, Jr. of the Santa Barbara Research Center headed the infrared radiometer experiment.[32]

**Ultraviolet spectrometers**

Two ultraviolet spectrometers were involved in this experiment, one to measure UV absorption, the other to sense UV emissions. The occultation spectrometer scanned Mercury's edge as it passed in front of the Sun, and detected whether solar ultraviolet radiation was absorbed at certain wavelengths, which would indicate the presence of gas particles, and therefore an atmosphere.[35] The airglow spectrometer detected extreme ultraviolet radiation emanating from atoms of gaseous hydrogen, helium, carbon, oxygen, neon, and argon.[36][32] Unlike the occultation spectrometer, it did not require backlighting from the Sun, and could move along with the rotatable scan platform on the spacecraft. The experiment's most important goal was to ascertain whether Mercury had an atmosphere, but would also gather data at Earth and Venus and study the interstellar background radiation.[34]

**Plasma detectors**

The plasma experiment's goal was to study the ionized gases (plasma) of the solar wind, the temperature and density of its electrons, and how the planets affected the velocity of the plasma stream.[37] The experiment contained two components, facing in opposite directions. The Scanning Electrostatic Analyzer was aimed toward the Sun, and could detect positive ions and electrons, which were separated by a set of three concentric hemispherical plates. The Scanning Electron Spectrometer was aimed away from the Sun, and detected only electrons, using just one hemispherical plate. The instruments could be rotated about 60 degrees to either side.[32] By gathering data on the solar wind's movement around Mercury, the plasma experiment could be used to verify the magnetometer's observations of Mercury's magnetic field.[34] Using the plasma detectors, Mariner 10 gathered the first *in situ* solar wind data from inside Venus' orbit.[38]

Shortly after launch, scientists found that the Scanning Electrostatic Analyzer had failed because a door shielding the analyzer did not open. An unsuccessful attempt was made to forcibly unfasten the door with the first course correction maneuver.[39] The experiment operators had planned to observe the directions taken by positive ions prior to the ions' collision with the Analyzer, but this data was lost.[40] The experiment was still able to collect some data using the properly functioning Scanning Electron Spectrometer.[41]

**Charged particle telescopes**

The goal of the charged particles experiment was to observe how the heliosphere interacted with cosmic radiation.[*][42] In connection with the plasma detectors and magnetometers, this experiment had the potential to provide additional evidence of a magnetic field around Mercury,[*][43] by showing whether such a field had captured charged particles.[*][32] Two telescopes were used to collect highly energetic electrons and atomic nuclei, specifically oxygen nuclei or less massive.[*][44] These particles then passed through a set of detectors and were counted.[*][32]

**Magnetometers**

Two fluxgate magnetometers were entrusted with discerning whether Mercury produced a magnetic field,[*][45] and studying the interplanetary magnetic field between flybys.[*][46] In designing this experiment, scientists had to account for interference from the magnetic field generated by Mariner 10's many electronic parts. For this reason, the magnetometers had to be situated on a long boom, one closer to the octagonal hub, the other one further away. Data from the two magnetometers would be cross-referenced to filter out the spacecraft's own magnetic field.[*][47][*][48] Drastically weakening the probe's magnetic field would have increased costs.[*][15]

**Celestial Mechanics and Radio Science experiment**

## 1.11.3 Departing the Earth–Moon system

*Mariner 10 imaged the Earth and Moon shortly after launch*

Boeing finished building the spacecraft at the end of June 1973, and Mariner 10 was delivered from Seattle to JPL's headquarters in California, where JPL comprehensively tested the integrity of the spacecraft and its instruments.

After the tests were finished, the probe was transported to the Eastern Test Range in Florida, the launch site. Technicians filled a tank on the spacecraft with 29 kilograms (64 lb) of hydrazine fuel so that the probe could make course corrections, and attached squibs, whose detonation would signal Mariner 10 to exit the launch rocket and deploy its instruments.[*][49][*][50] The planned gravity assist at Venus made it feasible to use an Atlas-Centaur rocket instead of a more powerful but more expensive Titan IIIC.[*][14][*][51] The probe and the Atlas-Centaur were attached together ten days prior to liftoff. Launch posed one of the largest risks of failure for the Mariner 10 mission—Mariner 1, Mariner 3, and Mariner 8 all failed minutes after lift-off due to either engineering failures or Atlas rocket malfunctions.[*][24][*][52][*][53] The mission had a launch window of about a month in length, from October 16, 1973, to November 21. NASA chose November 3 as the launch date because it would optimize imaging conditions when the spacecraft arrived at Mercury.[*][51]

*Launch of Mariner 10*

On November 3 at 12:45 am Eastern Time, the Atlas-Centaur carrying Mariner 10 lifted off from pad SLC-36B. The Atlas stage burned for around four minutes, after which it was jettisoned, and the Centaur stage took over for an additional five minutes, propelling Mariner 10 to a parking orbit. The temporary orbit took the spacecraft one-third of the distance around Earth: this maneuver was needed to reach the correct spot for a second burn by the Centaur engines, which set Mariner 10 on a path towards Venus. The probe then separated from the rocket; subsequently, the Centaur stage diverted away to avoid the possibility of future collision. Never before had a planetary mission depended upon two separate rocket burns during the launch, and even with Mariner 10, scientists initially viewed the maneuver as too risky.[54][55]

During its first week of flight, the Mariner 10 camera system was tested by taking five photographic mosaics of the Earth and six of the Moon. It also obtained photographs of the north polar region of the Moon where prior coverage was poor. These photographs provided a basis for cartographers to update lunar maps and improve the lunar control net.[56]

## 1.11.4   Cruise to Venus

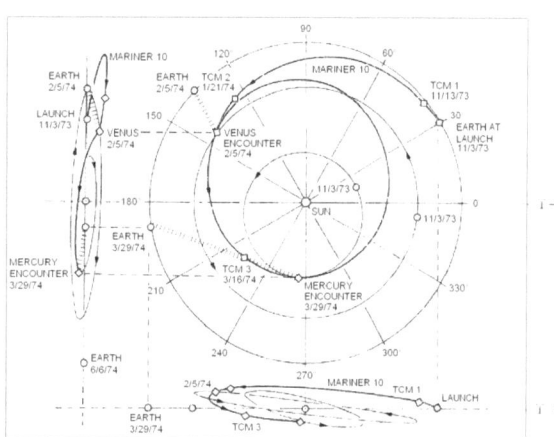

*Trajectory of Mariner 10 spacecraft: since launch on November 3, 1973, to first fly-by of Mercury on March 29, 1974*

Far from being an uneventful cruise, Mariner 10's three month journey to Venus was fraught with technical malfunctions, which kept mission control on edge.[57] Donna Shirley recounted her team's frustration: "It seemed as if we were always just patching Mariner 10 together long enough to get it on to the next phase and next crisis."[58] A trajectory correction maneuver was made on November 13, 1973. Immediately afterwards, the star-tracker locked onto a bright flake of paint which had come off the spacecraft and lost tracking on the guide star Canopus. An automated safety protocol recovered Canopus, but the problem of flaking paint recurred throughout the mission. The on-board computer also experienced unscheduled resets occasionally, which necessitated reconfiguring the clock sequence and subsystems. Periodic problems with the high-

gain antenna also occurred during the cruise.

In January 1974, Mariner 10 made ultraviolet observations of Comet Kohoutek. Another mid-course correction was made on January 21, 1974.

## 1.11.5   Venus flyby

The spacecraft passed Venus on February 5, 1974, the closest approach being 5,768 km at 17:01 UT. It was the twelfth spacecraft to reach Venus and the eighth to return data from the planet,[59] as well as the first mission to successfully broadcast images of Venus back to Earth.[60] As Mariner 10 veered around Venus, from the planet's night side to daylight, the cameras snapped the probe's first image of Venus, showing an illuminated arc of clouds over the north pole emerging from darkness. Engineers initially feared that the star-tracker could mistake the much brighter Venus for Canopus, repeating the mishaps with flaking paint. Fortunately, the star-tracker did not malfunction. Earth occultation occurred between 17:07 UT and 17:11 UT, during which the spacecraft transmitted X-band radio waves through Venus' atmosphere, gathering data on cloud structure and temperature.[61][62] Although Venus's cloud cover is nearly featureless in visible light, it was discovered that extensive cloud detail could be seen through Mariner's ultraviolet camera filters. Earth-based ultra-violet observation had shown some indistinct blotching even before Mariner 10, but the detail seen by Mariner was a surprise to most researchers. The probe continued photographing Venus until February 13.[63] Among the encounter's 4,165 acquired photographs, one resulting series of images captured a thick and distinctly patterned atmosphere making a full revolution every four days,[64] just as terrestrial observations had suggested.[65]

The gravity assist was also a success, coming well within the acceptable margin for error. In the four hours between 16:00 UT and 20:00 UT on February 5, Mariner 10's heliocentric velocity dropped from 82,785 mph to 72,215 mph.[66] This changed the shape of the spacecraft's elliptical orbit around the Sun,[60] so that the perihelion now coincided with the orbit of Mercury.[66]

- Venus encounter

- Venus in real colors, processed from clear and blue filtered Mariner 10 images

- Mariner photograph of Venus in ultraviolet light

## 1.11.6   First Mercury flyby

The spacecraft flew past Mercury three times. The first Mercury encounter took place at 20:47 UT on March 29,

1974, at a range of 703 kilometers (437 mi), passing on the shadow side.*[3]

- First Mercury encounter

- 6 hours before closest approach

- 6 hours after closest approach

### 1.11.7 Second Mercury flyby

After looping once around the Sun while Mercury completed two orbits, Mariner 10 flew by Mercury again on September 21, 1974, at a more distant range of 48,069 km (29,869 mi) below the southern hemisphere.*[3]

- Second Mercury encounter

- Mosaic of images from the second encounter, covering the equator to the south pole

### 1.11.8 Third Mercury flyby

After losing roll control in October 1974, a third and final encounter, the closest to Mercury, took place on March 16, 1975, at a range of 327 km (203 mi), passing almost over the north pole.*[3]

- Third Mercury encounter

- Mercury in color

- Mercury in black and white

- Mercury in false-color

- A prominent scarp, Discovery Rupes, photographed during first flyby

- Representation of the thrust fault at Discovery Rupes

- Australia region

- Aurora region

- Caduceata region

- Old basin, 190 km in diameter, filled by smooth plains. The basin's hummocky rim is partly degraded and cratered by later events

### 1.11.9 End of mission

With its maneuvering gas just about exhausted, Mariner 10 started another orbit of the Sun. Engineering tests were continued until March 24, 1975,*[3] when final depletion of the nitrogen supply was signaled by the onset of an unprogrammed pitch turn. Commands were sent immediately to the spacecraft to turn off its transmitter, and radio signals to Earth ceased. Mariner 10 is still orbiting the Sun, although its electronics have probably been damaged by the Sun's radiation.*[67]

### 1.11.10 Discoveries

During its flyby of Venus, Mariner 10 discovered evidence of rotating clouds and a very weak magnetic field. Using a near-ultraviolet filter, it photographed Venus's chevron clouds and performed other atmospheric studies.

The spacecraft flew past Mercury three times. Owing to the geometry of its orbit – its orbital period was almost exactly twice Mercury's – the same side of Mercury was sunlit each time, so it was only able to map 40–45% of Mercury's surface, taking over 2,800 photos. It revealed a more or less Moon-like surface. It thus contributed enormously to our understanding of Mercury, whose surface had not been successfully resolved through telescopic observation. The regions mapped included most or all of the Shakespeare, Beethoven, Kuiper, Michelangelo, Tolstoj, and Discovery quadrangles, half of Bach and Victoria quadrangles, and small portions of Solitudo Persephones (later Neruda), Liguria (later Raditladi), and Borealis quadrangles.*[68]

Mariner 10 also discovered that Mercury has a tenuous atmosphere consisting primarily of helium, as well as a magnetic field and a large iron-rich core. Its radiometer readings suggested that Mercury has a night time temperature of −183 °C (−297 °F) and maximum daytime temperatures of 187 °C (369 °F).

Planning for MESSENGER, a spacecraft that surveyed Mercury until 2015, relied extensively on data and information collected by Mariner 10.

### 1.11.11 Mariner 10 Commemoration

In 1975, the US Post Office issued a commemorative stamp featuring the Mariner 10 space probe. The 10-cent Mariner 10 commemorative stamp was issued on April 4, 1975, at Pasadena, California.*[69]

### 1.11.12 See also

- 1973 in spaceflight

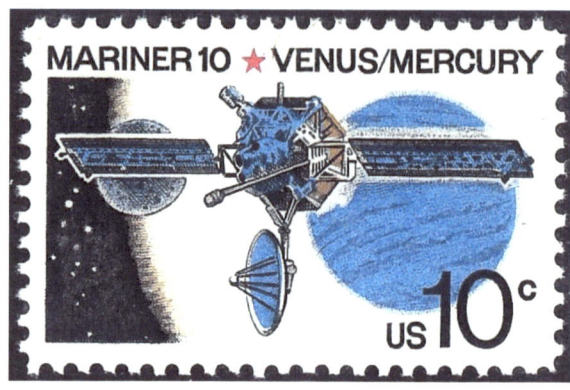

*Mariner 10 Space probe, Issue of 1975*

- Timeline of artificial satellites and space probes

## 1.11.13   References

**Notes**

[1] "Mariner 10". *National Space Science Data Center*. National Aeronautics and Space Administration. Retrieved 7 September 2013.

[2] Schudel, Matt (30 August 2013). "Bruce C. Murray, NASA space scientist, dies at 81". *The Washington Post*. Retrieved 31 August 2013.

[3] "Mariner 10". Retrieved 2 February 2014.

[4] Clark 2007, pp. 22–23

[5] Strom and Sprague 2003, pp. 16

[6] "Mariner 10 Quicklook". Retrieved 31 July 2014.

[7] Reeves 1994, pp. 222

[8] Biggs, John R.; Downhower, Walter J. (June 1974), "Mariner Venus/Mercury '73: A Strategy of Cost Control", *Astronautics & Aeronautics* (New York: The American Institute of Aeronautics and Astronautics) **12** (5): 48–53

[9] Murray and Burgess 1977, pp. 142

[10] Doody, Dave (29 October 2013). "Chapter 11. Typical Onboard Systems". *The Basics of Space Flight*. Jet Propulsion Laboratory. Retrieved 24 July 2015.

[11] Dunne and Burgess 1977, pp. 58

[12] Murray and Burgess 1977, pp. 50

[13] Ezell, Edward Clinton; Ezell, Linda Neuman (2009). *On Mars: Exploration of the Red Planet 1958-1978*. Mineola: Dover Publications. p. 445.

[14] Paul, Floyd A. (January 15, 1976). "Technical Memorandum 33-759: A Study of Mariner 10 Flight Experiences and Some Flight Piece Part Failure Rate Computations" (PDF). Jet Propulsion Laboratory. Retrieved 23 June 2015.

[15] Shirley, Donna L. (2003). "The Mariner 10 Mission to Venus and Mercury". *Acta Astronautica* (International Academy of Astronautics) **53** (4–10): 375–385. Retrieved 24 July 2015.

[16] Wilson, James H. (15 October 1973). "Technical Memorandum 33-657: Mariner Venus Mercury 1973" (PDF). Pasadena: Jet Propulsion Laboratory. p. 12. Retrieved 8 September 2015.

[17] Dunne and Burgess 1978, pp. 32–33

[18] Murray and Burgess 1977, pp. 21

[19] Dunne and Burgess 1978, pp. 30–32

[20] Reeves 1994, pp. 242

[21] Dunne and Burgess 1978, pp. 56

[22] Murray and Burgess 1977, pp. 25–26

[23] Strom and Sprague 2003, pp. 14

[24] Murray and Burgess 1977, pp. 38

[25] Dunne and Burgess 1978, pp.19

[26] Giberson and Cunningham 1975, pp. 719

[27] NASA/NSSDC – Mariner 10 – Television Photography

[28] Murray and Burgess 1977, pp. 43–48

[29] Clark, Pamela, ed. (December 2003). "Mariner 10: A Retrospective" (PDF). *Mercury Messenger* (Lunar and Planetary Institute) (10). Retrieved 25 May 2015.

[30] "Bulletin No. 14: TCM-2 Performance Superb TV Heaters Have Come On" (PDF). Mariner Venus/Mercury 1973 Project Office. 23 January 1974. Retrieved 25 May 2015.

[31] Dunne and Burgess 1978, pp. 57–58

[32] *Science Instrument Survey*. Moffett Field: Ames Research Center, NASA. May 1973. pp. 148–167.

[33] Dunne and Burgess 1978, pp. 21-22

[34] Strom and Sprague 2003, pp. 18-19

[35] Dunne and Burgess 1978, pp. 25-26

[36] Rothery 2015, pp. 26

[37] "Scanning Electrostatic Analyzer and Electron Spectrometer". *National Space Science Data Center*. National Aeronautics and Space Administration. Retrieved 27 July 2015.

[38] Dunne and Burgess 1978, pp. 22-23

[39] "Bulletin No. 7: First Trajectory Correction Maneuver A Success" (PDF). Mariner Venus/Mercury 1973 Project Office. 13 November 1973. Retrieved 25 May 2015. line feed character in |title= at position 53 (help)

[40] "Bulletin No. 15: Venus Flyby Set For Tuesday at 10:01 A.M. PT" (PDF). Mariner Venus/Mercury 1973 Project Office. 1 February 1974. Retrieved 7 September 2015.

[41] Dunne and Burgess 1978, pp. 47

[42] Strom and Sprague 2003, pp. 19

[43] Rothery 2015, pp. 28

[44] Dunne and Burgess 1978, pp. 24

[45] Rothery 2015, pp. 27

[46] Dunne and Burgess 1978, pp. 24

[47] Murray and Burgess 1977, pp. 95

[48] Strom and Sprague 2003, pp. 19

[49] Dunne and Burgess 1978, pp.42

[50] Murray and Burgess 1977, pp. 36–37

[51] Strom and Sprague 2003, pp. 14–16

[52] "Mariner 1". *National Space Science Data Center*. National Aeronautics and Space Administration. Retrieved 22 August 2015.

[53] "Mariner 3 Failure Laid to Shroud", *The Spokesman-Review*, 13 November 1964, p. 21, retrieved 22 August 2015

[54] Bowles, Mark D. (2004). *Taming Liquid Hydrogen: The Centaur Upper Stage Rocket 1958-2002*. Washington D.C.: Government Printing Office. pp. 131–133.

[55] Dunne and Burgess 1977, pp. 45-46

[56] Dunne and Burgess 1978, pp. 47–53.

[57] Murray and Burgess 1977, pp. 55

[58] Shirley 1998, pp. 91

[59] Williams, David R. (29 May 2014). "Chronology of Venus Exploration". *National Space Science Data Center*. National Aeronautics and Space Administration. Retrieved 8 September 2015.

[60] Ulivi and Harland 2007, pp. 181

[61] Murray and Burgess 1977, pp. 61-64

[62] Dunne and Burgess 1978, pp. 61-63

[63] Murray and Burgess 1977, pp. 79

[64] Reeves 1994, pp. 244

[65] Dunne and Burgess 1978, pp. 68

[66] "Bulletin No. 18: Mariner 10 Enroute to Mercury - Continues Query of Venus" (PDF). Mariner Venus/Mercury 1973 Project Office. 6 February 1974. Retrieved 7 September 2015. line feed character in |title= at position 59 (help)

[67] Mariner 10 (2006) *Views of the Solar System*

[68] Schaber, Gerald G.; McCauley, John F. *Geologic Map of the Tolstoj (H-8) Quadrangle of Mercury* (PDF). U.S. Geological Survey. USGS Miscellaneous Investigations Series Map I–1199, as part of the Atlas of Mercury, 1:5,000,000 Geologic Series. Retrieved 12 November 2007.

[69] Piazza, Jill (September 8, 2008). "10-cent Mariner 10". *Arago*. Smithsonian National Postal Museum. Retrieved 22 August 2015.

**Further reading**

- Clark, Pamela Elizabeth (2007). *Dynamic Planet: Mercury in the Context of its Environment*. New York: Springer Science+Business Media, LLC.

- Dunne, James A.; Burgess, Eric (1978). *The Voyage of Mariner 10: Mission to Venus and Mercury (NASA SP-424)*. Washington, D.C.: National Aeronautics and Space Administration Scientific and Technical Information Office.

- Giberson, W. Eugene; Cunningham, N. William (4 February 1975). "Mariner 10 Mission to Venus and Mercury". *Acta Astronautica* (Pergamon Press) **2**: 715–743.

- Murray, Bruce; Burgess, Eric (1977). *Flight to Mercury*. New York: Columbia University Press.

- Reeves, Robert (1994). *The Superpower Space Race: An Explosive Rivalry Through the Solar System*. New York: Plenum Press.

- Rothery, David A. (2015). *Planet Mercury: From Pale Pink Dot to Dynamic World*. Cham: Springer International Publishing.

- Shirley, Donna (1998). *Managing Martians*. New York: Broadway Books.

- Strom, Robert G.; Sprague, Ann L. (2003). *Exploring Mercury: The Iron Planet*. Chichester: Praxis Publishing Ltd.

- Ulivi, Paolo; Harland, David M. (2007). *Part 1: The Golden Age 1957-1982*. Robotic Exploration of the Solar System (Chichester ed.). Praxis Publishing Ltd.

### 1.11.14 External links

- *The Voyage of Mariner 10: Mission to Venus and Mercury* (NASA SP-424) 1978 This is an *entire book* about Mariner 10, with all pictures and diagrams, on-line! Scroll down to click on the "Table of Contents" link. PDF version

- 'Mariner 10', NASA's 1973–75 Venus/Mercury Mission

- Mariner 10 image archive

- Mariner 10 mission bulletins

- Mariner 10 Mission Profile by NASA's Solar System Exploration

- Calibrated images from the Mariner 10 mission to Mercury and Venus

- Master Catalog entry for Mariner 10 at the National Space Science Data Center

- Boeing: History – Products – Boeing Mariner 10 Spacecraft

## 1.12 Mariner 2

**Mariner 2** (**Mariner-Venus 1962**), an American space probe to Venus, was the first robotic space probe to conduct a successful planetary encounter. The first successful spacecraft in the NASA Mariner program, it was a simplified version of the Block I spacecraft of the Ranger program and an exact copy of Mariner 1. The missions of Mariner 1 and 2 spacecraft are together sometimes known as the Mariner R missions. The probes had originally been intended to launch on the Atlas-Centaur, but that vehicle was not ready in time for the missions, forcing NASA to use the Atlas-Agena which had a much smaller payload capacity and necessitated greatly simplified probes with a minimum of instrumentation. The Mariner 2 spacecraft was launched from Cape Canaveral on August 27, 1962 and passed as close as 34,773 kilometers (21,607 mi) to Venus on December 14, 1962.[2]

The Mariner probe consisted of a 100 cm (39.4 in) diameter hexagonal bus, to which solar panels, instrument booms, and antennas were attached. The scientific instruments on board the Mariner spacecraft were two radiometers (one each for the microwave and infrared portions of the spectrum), a micrometeorite sensor, a solar plasma sensor, a charged particle sensor, and a magnetometer. These instruments were designed to measure the temperature distribution on the surface of Venus, as well as making basic measurements of Venus' atmosphere.

The primary mission was to receive communications from the spacecraft in the vicinity of Venus and to perform radiometric temperature measurements of the planet. A second objective was to measure the interplanetary magnetic field and charged particle environment.[3][4]

En route to Venus, Mariner 2 measured the solar wind, a constant stream of charged particles flowing outwards from the Sun, confirming the measurements by Luna 1 in 1959. It also measured interplanetary dust, which turned out to be scarcer than predicted. In addition, Mariner 2 detected high-energy charged particles coming from the Sun, including several brief solar flares, as well as cosmic rays from outside the Solar System. As it flew by Venus on December 14, 1962, Mariner 2 scanned the planet with its pair of radiometers, revealing that Venus has cool clouds and an extremely hot surface.

### 1.12.1 Spacecraft and subsystems

The Mariner 2 spacecraft was designed and built by the Jet Propulsion Laboratory of the California Institute of Technology.[5] It consisted of a hexagonal base, 1.04 meters across and 0.36 meters thick, which contained six magnesium chassis housing the electronics for the science experiments, communications, data encoding, computing, timing, and attitude control, and the power control, battery, and battery charger, as well as the attitude control gas bottles and the rocket engine. On top of the base was a tall pyramid-shaped mast on which the science experiments were mounted, which brought the total height of the spacecraft to 3.66 meters. Attached to either side of the base were rectangular solar panel wings with a total span of 5.05 meters and width of 0.76 meters. Attached by an arm to one side of the base and extending below the spacecraft was a large directional dish antenna.

The power system of Mariner 2 consisted of two solar cell wings, one 183 cm by 76 cm and the other 152 cm by 76 cm (with a 31 cm dacron extension (a solar sail) to balance the solar pressure on the panels), which powered the craft directly or recharged a 1000 watt-hour sealed silver-zinc cell battery. This battery was used before the panels were deployed, when the panels were not illuminated by the Sun, and when loads were heavy. A power-switching and booster regulator device controlled the power flow. Communications consisted of a 3-watt transmitter capable of continuous telemetry operation, the large high gain directional dish antenna, a cylindrical omnidirectional antenna at the top of the instrument mast, and two command antennas, one on the end of either solar panel, which received instructions for midcourse maneuvers and other functions.

Propulsion for midcourse maneuvers was supplied by a monopropellant (anhydrous hydrazine) 225 N retro-rocket.

*Launch of Mariner 2*

The hydrazine was ignited using nitrogen tetroxide and aluminum oxide pellets, and thrust direction was controlled by four jet vanes situated below the thrust chamber. Attitude control with a 1 degree pointing error was maintained by a system of nitrogen gas jets. The Sun and Earth were used as references for attitude stabilization. Overall timing and control was performed by a digital Central Computer and Sequencer. Thermal control was achieved through the use of passive reflecting and absorbing surfaces, thermal shields, and movable louvers.

### Scientific instruments

Only 40 pounds (18 kg) of the spacecraft could be allocated to scientific experiments.[*][6] The following scientific instruments were mounted on the instrument mast and base:

- A two-channel **microwave radiometer** of the crystal video type operating in the standard Dicke mode of chopping between the main antenna, pointed at the target, and a reference horn pointed at cold space.[*][7] It was used to determine the absolute temperature of Venus' surface and details concerning its atmosphere through its microwave-radiation characteristics, including the daylight and dark hemispheres, and in the region of the terminator. Measurements were performed simultaneously in two frequency bands of 13.5 mm and 19 mm.[*][6][*][8] The total weight of the radiometer was 22 pounds (10 kg). Its average power consumption was 4 watts and its peak power consumption 9 watts.[*][9]

- A two-channel **infrared radiometer** to measure the effective temperatures of small areas of Venus. The radiation that was received could originate from the planetary surface, clouds in the atmosphere, the atmosphere itself or a combination of these. The radiation was received in two spectral ranges: 8 to 9 μm (0.00031 to 0.00035 inches) (focused on 8.4 μm) and 10 to 10.8 μm (0.00039 to 0.00043 inches) (focused on 10.4 μm).[*][6] The latter corresponding to the carbon dioxide band.[*][10] The total weight of the infrared radiometer, which was housed in a magnesium casting, was 1.3 kg, and it required 2.4 watts of power. It was designed to measure radiation temperatures between 200 and approximately 500 K.[*][11]

- A three-axis **fluxgate magnetometer** to measure planetary and interplanetary magnetic fields.[*][6] Three probes were incorporated in its sensors, so it could obtain three mutually orthogonal components of the field vector. Readings of these components were separated by 1.9 seconds. It had three analog outputs that had each two sensitivity scales: ± 64 γ and ± 320 γ (1 γ = 1 nanotesla). These scales were automatically switched by the instrument. The field that the magnetometer observed was the super-position of a nearly constant spacecraft field and the interplanetary field. Thus, it effectively measured only the changes in the interplanetary field.[*][12]

- An **ionization chamber** with matched **Geiger-Müller tubes** (also known as a cosmic ray detector) to measure high-energy cosmic radiation.[*][6][*][13]

- A **particle detector** (implemented through use of an Anton type 213 Geiger-Müller tube) to measure lower radiation (especially near Venus),[*][6][*][14] also known as the Iowa detector, as it was provided by the University of Iowa.[*][13] It was a miniature tube having a 1.2 mg/cm² mica window about 0.3 cm in diameter and weighing about 60 g. It detected soft x-rays efficiently and ultraviolet inefficiently, and was previously used in Injun 1, Explorer 12 and Explorer 14.[*][14] It was able to detect protons above 500 keV in energy and electrons above 35 keV.[*][3] The length of the basic telemetry frame was 887.04 seconds. During each frame, the counting rate of the detector was sampled twice at intervals separated by 37 seconds. The first sampling was the number of counts during an interval of 9.60 seconds (known as the 'long gate'); the second was the number of counts during an interval of 0.827 seconds (known as the 'short gate'). The long gate accumulator overflowed on the 256th count and the short gate accumulator overflowed on the 65,536th count. The maximum counting rate of the tube was 50,000 per second.[*][14]

- A **cosmic dust detector** to measure the flux of cosmic dust particles in space.[*][6]

- A **solar plasma spectrometer** to measure the spectrum of low-energy positively charged particles from the Sun, i.e. the solar wind.[*][6]

The magnetometer was attached to the top of the mast below the omnidirectional antenna. Particle detectors were mounted halfway up the mast, along with the cosmic ray detector. The cosmic dust detector and solar plasma spectrometer were attached to the top edges of the spacecraft base. The microwave radiometer, the infrared radiometer and the radiometer reference horns were rigidly mounted to a 48 cm diameter parabolic radiometer antenna mounted near the bottom of the mast. All instruments were operated throughout the cruise and encounter modes except the radiometers, which were only used in the immediate vicinity of Venus.

In addition to these scientific instruments, Mariner 2 had a data conditioning system (DCS) and a scientific power switching (SPS) unit. The DCS was a solid-state electronic system designed to gather information from the scientific instruments on board the spacecraft. It had four basic functions: analog-to-digital conversion, digital-to-digital conversion, sampling and instrument-calibration timing, and planetary acquisition. The SPS unit was designed to perform the following three functions: control of the application of AC power to appropriate portions of the science subsystem, application of power to the radiometers and removal of power from the cruise experiments during radiometer calibration periods, and control of the speed and direction of the radiometer scans. The DCS sent signals to the SPS unit to perform the latter two functions.[*][6]

### Mission objectives

The scientific objectives were:[*][3]

- Radiometer experiment.

- Infrared experiment.

- Magnetometer experiment.

- Charged particles experiment.

- Plasma experiment.

- Micrometeorite experiment.

Besides the experiments with the scientific instruments, the objectives of both the Mariner 1 and 2 probes included also engineering objectives:[*][3]

- Evaluation of the attitude control system.

- Evaluation of the environmental control system.

- Evaluation of the entire power system.

- Evaluation of the communication system.

## 1.12.2   Mission profile

### Launch

*Mariner Atlas-Agena ignition*

Mariner 2 was launched from Cape Canaveral Air Force Station Launch Complex 12 at 06:53:14 UTC on August 27, 1962 by a two-stage Atlas-Agena rocket.[*][6][*][15] The two-stage Atlas-Agena rocket carrying Mariner 1 had veered off-course during its launch on July 22, 1962 due to a defective signal from the Atlas and a bug in the program equations of the ground-based guiding computer, and subsequently the spacecraft was destroyed by the Range Safety Officer. Mariner 2 nearly met the same fate as its predecessor when one of the Atlas's verniers moved to maximum stop shortly before booster engine cutoff. This caused a rapid roll of the launch vehicle that quickly approached one revolution per second. With the structural integrity of the booster in jeopardy, the range safety officer prepared to issue the destruct command, but almost as soon as it started, the rolling motion stopped and the launch proceeded uneventfully. The incident was traced to a loose wire in the guidance computer which was pushed back into place by the centrifugal force of the roll.

Five minutes after lift-off, the Atlas and Agena-Mariner separated, followed by the first Agena burn and second Agena burn. The Agena-Mariner separation injected the Mariner 2 spacecraft into a geocentric escape hyperbola at 26 minutes 3 seconds after lift-off. The NASA NDIF tracking station at Johannesburg, South Africa, acquired the spacecraft about 31 minutes after launch. Solar panel extension was completed approximately 44 minutes after launch. The Sun lock acquired the Sun about 18 minutes later. The high-gain antenna was extended to its acquisition angle of 72°. The output of the solar panels was slightly above the predicted output. As all subsystems were performing normally, as the battery was fully charged, and the solar panels were providing adequate power, the decision was made on August 29 to turn on cruise science experiments. On September 3, the Earth acquisition sequence was initiated, and Earth lock was established 29 minutes later.[*][6]

**Mid-course maneuver**

The accuracy of the Atlas-Agena was such that a mid-course correction was required to satisfy the mission requirements. The mid-course correction consisted of a roll-turn sequence, followed by a pitch-turn sequence and finally a motor-burn sequence. Preparation commands were sent to the spacecraft at 21:30 UTC on September 4. Initiation of the mid-course maneuver sequence was sent at 22:49:42 UTC and the roll-turn sequence started one hour later. The entire maneuver took approximately 34 minutes.[*][6]

Due to the mid-course maneuver, the sensors lost their lock with the Sun and Earth. At 00:27:00 UTC the Sun re-acquisition begun and at 00:34 UTC the Sun was reacquired. Earth re-acquisition started at 02:07:29 UTC and Earth was reacquired at 02:34 UTC.[*][6]

**Loss of attitude control**

On September 8 at 12:50 UTC, the spacecraft experienced a problem with attitude control. It automatically turned on the gyros, and the cruise science experiments were automatically turned off. The exact cause is unknown as attitude sensors went back to normal before telemetry measurements could be sampled, but it may have been an Earth-sensor malfunction or a collision with a small unidentified

object which temporarily caused the spacecraft to lose Sun lock. A similar experience happened on September 29 at 14:34 UTC. Again, all sensors went back to normal before it could be determined which axis had lost lock. By this date, the Earth sensor brightness indication had essentially gone to zero. This time, however, telemetry data indicated that the Earth-brightness measurement had increased to the nominal value for that point in the trajectory.[*][6]

**Solar panel output**

On October 31, the output from one solar panel (with solar sail attached) deteriorated abruptly. It was diagnosed as a partial short circuit in the panel. As a precaution, the cruise science instruments were turned off. A week later, the panel resumed normal function, and cruise science instruments were turned back on. The panel permanently failed on November 15, but Mariner 2 was close enough to the Sun that one panel could supply adequate power; thus, the cruise science experiments were left active.[*][6]

**Encounter with Venus**

Mariner 2 was the first spacecraft to successfully encounter another planet, passing as close as 34,773 kilometers (21,607 mi) to Venus on December 14, 1962.[*][2]

**Post encounter**

After encounter, cruise mode resumed. Spacecraft perihelion occurred on December 27 at a distance of 105,464,560 km. The last transmission from Mariner 2 was received on January 3, 1963 at 07:00 UTC, making the total time from launch to termination of the Mariner 2 mission 129 days.[*][16] Mariner 2 remains in heliocentric orbit.

### 1.12.3 Results

The data produced during the flight consisted of two categories, namely tracking data and telemetry data.[*][16]

**Scientific observations**

The microwave radiometer made three scans of Venus in 35 minutes on December 14, 1962 starting at 18:59 UTC.[*][9] The first scan was made on the dark side, the second was near the terminator, and the third was located on the light side.[*][9][*][17] The scans with the 19 mm band revealed peak temperatures of 490 ± 11 K on the dark side, 595 ± 12 K near the terminator, and 511 ± 14 K on the light side.[*][18] It was concluded that there is no significant difference in temperature across Venus.[*][9][*][17] However, the results suggest a limb darkening, an effect which presents cooler temperatures near the edge of the planetary disk and higher temperatures near the center.[*][7][*][8][*][9][*][17][*][18][*][19] This also supported the theory that the Venusian surface was extremely hot or the atmosphere optically thick.[*][9][*][17][*][18]

The infrared radiometer showed that the 8.4 μm and 10.4 μm radiation temperatures were in agreement with radiation temperatures obtained from Earth-based measurements.[*][11] There was no systematic difference between the temperatures measured on the light side and dark side of the planet, which was also in agreement with Earth-based measurements.[*][11] The limb darkening effect that the microwave radiometer detected was also present in the measurements by both channels of the infrared radiometer.[*][11][*][17][*][19] The effect was only slightly present in the 10.4 μm channel but was more pronounced in the 8.4 μm channel.[*][17] The 8.4 μm channel also showed a slight phase effect. The phase effect indicated that if a greenhouse effect existed, heat was transported in an efficient manner from the light side to the dark side of the planet.[*][17] The 8.4 μm and 10.4 μm showed equal radiation temperatures, indicating that the limb darkening effect would appear to come from a cloud structure rather than the atmosphere.[*][11] Thus, if the measured temperatures were actually cloud temperatures instead of surface temperatures, then these clouds would have to be quite thick.[*][10][*][17][*][19]

The magnetometer detected a persistent interplanetary magnetic field varying between 2 γ and 10 γ, which agrees with prior Pioneer 5 observations from 1960. This also means that interplanetary space is rarely empty or field-free.[*][12] The magnetometer could detect changes of about 4 γ on any of the axes, but no trends above 10 γ were detected near Venus, nor were fluctuations seen like those that appear at Earth's magnetospheric termination. This means that Mariner 2 found no detectable magnetic field near Venus, although that didn't necessarily mean that Venus had none.[*][17][*][20] However, if Venus had a magnetic field, then it would have to be at least smaller than 1/10 the magnetic field of the Earth.[*][20][*][21] In 1980, the Pioneer Venus Orbiter indeed showed that Venus has a small weak magnetic field.[*][22]

The Anton type 213 Geiger-Müller tube performed as expected.[*][23] The average rate was 0.6 counts per second. Increases in its counting rate were larger and more frequent than for the two larger tubes, since it was more sensitive to particles of lower energy.[*][6] It detected 7 small solar bursts of radiation during September and October and 2 during November and December.[*][24] The absence of a detectable magnetosphere was also confirmed by the tube; it

detected no radiation belt at Venus similar to that of Earth. The count rate would have increased by $10^4$, but no change was measured.[*][6][*][25]

It was also shown that in interplanetary space, the solar wind streams continuously[*][15][*][26] and the cosmic dust density is much lower than the near-Earth region.[*][27] Improved estimates of Venus' mass and the value of the Astronomical Unit were made. Also, research, which was later confirmed by other explorations, suggested that Venus rotates very slowly and in a direction opposite that of the Earth.[*][28]

### 1.12.4   References

[1] McDowell, Jonathan. "Launch Log". *Jonathan's Space Page*. Retrieved 12 September 2013.

[2] "Mariner 2". US National Space Science Data Center. Retrieved 8 September 2013.

[3] Jet Propulsion Laboratory (under contract for NASA) (1962-06-15). "Tracking Information Memorandum No. 332-15: Mariner R 1 and 2" (PDF). California Institute of Technology. Retrieved 2008-01-24.

[4] Renzetti, N.A. (1965-07-01). "Technical Memorandum No. 33-212: Tracking and Data Acquisition Support for the Mariner Venus 1962 Mission" (PDF). NASA. Retrieved 2008-01-24.

[5] "The Mission of Mariner II: Preliminary Observations - Profile of Events". *Science, New Series* **138** (3545): 1095. 1962-12-07. Bibcode:1962Sci...138.1095.. doi:10.1126/science.138.3545.1095. PMID 17772964. (subscription required (help)).

[6] Jet Propulsion Laboratory (under contract for NASA) (July 1965). "Mariner-Venus 1962, Final Project Report" (PDF). California Institute of Technology. Retrieved 2008-01-27.

[7] Jones, Douglas E. (1966-01-01). "Technical Report No. 32-722: The Mariner II Microwave Radiometer Experiment" (PDF). Jet Propulsion Laboratory, California Institute of Technology. Retrieved 2009-02-15.

[8] Barath, F.T.; Barrett, A.H.; Copeland, J.; Jones, D.E.; Lilley, A.E. (February 1964). "Symposium on Radar and Radiometric Observations of Venus during the 1962 Conjunction: Mariner 2 Microwave Radiometer Experiment and Results". *The Astronomical Journal* **69** (1): 49–58. Bibcode:1964AJ.....69...49B. doi:10.1086/109227.

[9] Barath, F.T.; Barrett, A.H.; Copeland, J.; Jones, D.E.; Lilley, A.E. (1963-03-08). "Mariner II: Preliminary Reports on Measurements of Venus - Microwave Radiometers". *Science, New Series* **139** (3558): 908–909. Bibcode:1963Sci...139..908B. doi:10.1126/science.139.3558.908. PMID 17743052. (subscription required (help)).

[10] Chase, S.C.; Kaplan, L.D.; Neugebauer, G. (1963-03-08). "Mariner II: Preliminary Reports on Measurements of Venus - Infrared Radiometer". *Science, New Series* **139** (3558): 907–908. Bibcode:1963Sci...139..907C. doi:10.1126/science.139.3558.907. PMID 17743051. (subscription required (help)).

[11] Chase, S.C.; Kaplan, L.D.; Neugebauer, G. (1963-11-15). "The Mariner 2 Infrared Radiometer Experiment" (PDF). *Journal of Geophysical Research* **68** (22): 6157–6169. Bibcode:1963JGR....68.6157C. doi:10.1029/jz068i022p06157. Retrieved 2009-02-14.

[12] Coleman, Jr., Paul J.; Davis, Jr., Leverett; Smith, Edward J.; Sonett, Charles P. (1962-12-07). "The Mission of Mariner II: Preliminary Observations - Interplanetary Magnetic Fields". *Science, New Series* **138** (3545): 1099–1100. Bibcode:1962Sci...138.1099C. doi:10.1126/science.138.3545.1099. PMID 17772967. (subscription required (help)).

[13] Anderson, Hugh R. (1963-01-04). "Mariner II: High-Energy-Radiation Experiment". *Science, New Series* **139** (3549): 42–45. Bibcode:1963Sci...139...42A. doi:10.1126/science.139.3549.42. PMID 17752023. (subscription required (help)).

[14] Van Allen, James A.; Frank, Louis A. (1962-12-07). "The Mission of Mariner II: Preliminary Observations - The Iowa Radiation Experiment". *Science, New Series* **138** (3545): 1097–1098. Bibcode:1962Sci...138.1097V. doi:10.1126/science.138.3545.1097. PMID 17772965. (subscription required (help)).

[15] Neugebauer, M.; Snyder, C.W. (1962-12-07). "The Mission of Mariner II: Preliminary Observations - Solar Plasma Experiment". *Science, New Series* **138** (3545): 1095–1097. Bibcode:1962Sci...138.1095N. doi:10.1126/science.138.3545.1095-a. PMID 17772963. (subscription required (help)).

[16] Sparks, D.B. (March 1963). "The Mariner 2 Data Processing System" (fee required). California Institute of Technology. Retrieved 2008-01-28.

[17] Sonett, Charles P. (December 1963). "A Summary Review of the Scientific Findings of the Mariner Venus Mission". *Space Science Reviews* **2** (6): 751–777. Bibcode:1963SSRv....2..751S. doi:10.1007/BF00208814. (subscription required (help)).

[18] Pollack, James B.; Sagan, Carl (October 1967). "An Analysis of the Mariner 2 Microwave Observations of Venus". *The Astrophysical Journal* **150**: 327–344. Bibcode:1967ApJ...150..327P. doi:10.1086/149334.

[19] Kaplan, L.D. (June 1964). "Venus, Recent Physical Data for" (PDF). Retrieved 2009-02-15.

[20] Smith, Edward .J.; Davis, Jr., Leverett; Coleman, Jr., Paul J.; Sonett, Charles P. (1963-03-08). "Mariner II: Preliminary Reports on Measurements of Venus - Magnetic Field". *Science, New Series* **139** (3558): 909–910. Bibcode:1963Sci...139..909S. doi:10.1126/science.139.3558.909. PMID 17743053. (subscription required (help)).

[21] Smith, Edward J.; Davis, Jr., Leverett; Coleman, Jr., Paul J.; Sonett, Charles P. "Magnetic Measurements near Venus" (PDF). Retrieved 2009-02-15.

[22] Kivelson, Margaret G.; Russell, Christopher T. (1995). *Introduction to Space Physics.* Cambridge University Press. ISBN 978-0-521-45714-9.

[23] Van Allen, James A. (July 1964). "Survival of Thin Films in Space" (PDF). Department of Physics and Astronomy, State University of Iowa. Retrieved 2009-02-15.

[24] James, J.N. "Mariner II" (PDF). Retrieved 2009-02-15.

[25] Frank, L.A.; Van Allen, J.A.; Hills, H.K. (1963-03-08). "Mariner II: Preliminary Reports on Measurements of Venus - Charged Particles". *Science, New Series* **139** (3558): 905–907. Bibcode:1963Sci...139..905F. doi:10.1126/science.139.3558.905. PMID 17743050. (subscription required (help)).

[26] Ness, N.F.; Wilcox, J.M. (1964-10-12). "Solar Origin of the Interplanetary Magnetic Field". *Physical Review Letters* **13** (15): 461–464. Bibcode:1964PhRvL..13..461N. doi:10.1103/PhysRevLett.13.461. (subscription required (help)).

[27] Alexander, W.M. (1962-12-07). "The Mission of Mariner II: Preliminary Results - Cosmic Dust". *Science, New Series* **138** (3545): 1098–1099. Bibcode:1962Sci...138.1098A. doi:10.1126/science.138.3545.1098. PMID 17772966. (subscription required (help)).

[28] Goldstein, R.M.; Carpenter, R.L. (1963-03-08). "Rotation of Venus: Period Estimated from Radar Measurements". *Science, New Series* **139** (3558): 910–911. Bibcode:1963Sci...139..910G. doi:10.1126/science.139.3558.910. PMID 17743054. (subscription required (help)).

## 1.12.5   External links

- Mariner 2 Mission Profile by NASA's Solar System Exploration

- Full-scale engineering prototype of Mariner 2 in the Smithsonian Air and Space Museum, Washington, D.C.

- Mariner 2

*Launch of Mariner 5*

# 1.13   Mariner 5

**Mariner 5** (**Mariner Venus 1967**) was a spacecraft of the Mariner program that carried a complement of experiments to probe Venus' atmosphere by radio occultation, measure the hydrogen Lyman-alpha (hard ultraviolet) spectrum, and sample the solar particles and magnetic field fluctuations above the planet. Its goals were to measure interplanetary and Venusian magnetic fields, charged particles, plasma, radio refractivity and UV emissions of the Venusian atmosphere.

Mariner 5 was actually built as a backup to Mariner 4, but after the success of the Mariner 4 mission, it was modified for the Venus mission by removing the TV camera, reversing and reducing the four solar panels, and adding extra thermal insulation.

It was launched toward Venus on June 14, 1967 from Cape Canaveral Air Force Station Launch Complex 12 and flew by the planet on October 19 that year at an altitude of 3,990 kilometers (2,480 mi). With more sensitive instruments than its predecessor Mariner 2, Mariner 5 was able to shed new light on the hot, cloud-covered planet and on conditions in interplanetary space.

Radio occultation data from Mariner 5 helped to understand the temperature and pressure data returned by the Venera 4 lander, which arrived at Venus shortly before it. After these

missions, it was clear that Venus had a very hot surface and an atmosphere even denser than expected.

The operations of Mariner 5 ended in November 1967 and it is now defunct in a heliocentric orbit.

### 1.13.1   Further communication attempts

Further communication attempts were tried, in a joint spacecraft solar wind / solar magnetic fields investigation with Mariner 4, back in communication with Earth after being out of telemetry for about a year or more around superior conjunction. During the experiment, both spacecraft were going to be on the same idealized magnetic field spiral carried out from the sun by the solar wind.

Between April and November 1968 NASA tried to reacquire Mariner 5 to continue probing interplanetary conditions. Attempts to reacquire Mariner 5 during June, July, and early August 1968 yielded no spacecraft signal.

On October 14, the receiver operator at DSS 14 obtained a lock on the Mariner 5 signal. A carrier wave was detected, but outside expected frequency limits and varying in wavelength. Signal strength changes indicating the spacecraft was in a slow roll. Nevertheless, it was possible to lock the spacecraft to an uplink signal, but no response was observed to any commands sent to it. Without telemetry and without any signal change in response to commands, there was no possibility to repair or continue to use the spacecraft. Operations were terminated at the end of the track from DSS 61 at 07:46 GMT on November 5, 1968.

### 1.13.2   Instruments

1. Two-Frequency Beacon Receiver

2. S-Band Occultation

3. Helium Magnetometer

4. Interplanetary Ion Plasma Probe for E/Q of 40 to 9400 Volts

5. Celestial Mechanics

### 1.13.3   External links

- Mariner 5 Mission Profile by NASA's Solar System Exploration

- Mariner Venus 1967 Final Project Report

- The Mariner 5 flight path and its determination from tracking data

# 1.14 Mariner program

*Launch of Mariner 1 in 1962*

*This 1963 photo shows Dr. William H. Pickering, (center) JPL Director, presenting a Mariner 2 spacecraft model to President John F. Kennedy, (right). NASA Administrator James Webb is standing directly behind the Mariner model*

The **Mariner program** was a program conducted by the American space agency NASA in conjunction with Jet Propulsion Laboratory (JPL)[1] that launched a series of robotic interplanetary probes designed to investigate Mars, Venus and Mercury[2] from 1962 to 1973. The program included a number of firsts, including the first planetary flyby, the first pictures from another planet, the first planetary orbiter, and the first gravity assist maneuver.

Of the ten vehicles in the Mariner series, seven were successful. The planned Mariner Jupiter-Saturn vehicles evolved into Voyager 1 and Voyager 2 of the Voyager program,[3] while the Viking 1 and Viking 2 Mars orbiters were enlarged versions of the Mariner 9 spacecraft. Other Mariner-based spacecraft, launched since Voyager, included the Magellan probe to Venus, and the Galileo probe to Jupiter. A second-generation Mariner spacecraft, called the Mariner Mark II series, eventually evolved into the Cassini–Huygens probe, now in orbit around Saturn.

The total cost of the Mariner program was approximately $554 million.[4]

## 1.14.1 Basic layout

All Mariner spacecraft were based on a hexagonal or octagonal "bus", which housed all of the electronics, and to which all components were attached, such as antennae, cameras, propulsion, and power sources.[2][5] All of the Mariners launched after Mariner 2 had four solar panels for power, except for Mariner 10, which had two, and Mariner 2, which was based on the Ranger Lunar probe. Additionally, all except Mariner 1, Mariner 2 and Mariner 5 had TV cameras.

The first five Mariners were launched on Atlas-Agena rockets, while the last five used the Atlas-Centaur. All Mariner-based probes after Mariner 10 used the Titan IIIE, Titan IV unmanned rockets or the Space Shuttle with a solid-fueled Inertial Upper Stage and multiple planetary flybys.

## 1.14.2 Mariners 1 and 2

Main articles: Mariner 1 and Mariner 2

Mariner 1 (P-37) and Mariner 2 (P-38) were two deep-space probes making up NASA's Mariner-R project. The primary goal of the project was to develop and launch two spacecraft sequentially to the near vicinity of the planet Venus, receive communications from the spacecraft and to perform radiometric temperature measurements of the planet. A secondary objective was to make interplanetary magnetic field and/or particle measurements on the way to Venus and in the vicinity of Venus.[6][7] Mariner 1 (designated Mariner R-1) was launched on July 22, 1962, but

- Sensors: camera with digital tape recorder (about 20 pictures), cosmic dust, solar plasma, trapped radiation, cosmic rays, magnetic fields, radio occultation and celestial mechanics[9]

Status:

- Mariner 3 – Malfunctioned. Derelict in heliocentric orbit.[8]

- Mariner 4 – Communications lost after bombardment by micrometeoroids. Derelict in heliocentric orbit.[9]

### 1.14.4   Mariner 5

Main article: Mariner 5
The Mariner 5 spacecraft was launched to Venus on June

14, 1967 and arrived in the vicinity of the planet in October 1967. It carried a complement of experiments to probe Venus' atmosphere with radio waves, scan its brightness in ultraviolet light, and sample the solar particles and magnetic field fluctuations above the planet.

- Mission: Venus flyby

was destroyed approximately 5 minutes after liftoff by the Air Force Range Safety Officer when its malfunctioning Atlas-Agena rocket went off course. Mariner 2 (designated Mariner R-2) was launched on August 27, 1962, sending it on a 3½-month flight to Venus. The mission was a success, and Mariner 2 became the first spacecraft to have flown by another planet.

- Mission: Venus flyby

- weight: 203 kg (446 lb)

- Sensors: microwave and infrared radiometers, cosmic dust, solar plasma and high-energy radiation, magnetic fields

Status:

- Mariner 1 – Destroyed shortly after liftoff.

- Mariner 2 – Defunct after successful mission, occupies a heliocentric orbit.

### 1.14.3   Mariners 3 and 4

Main articles: Mariner 3 and Mariner 4
Mariner 3 and Mariner 4 were Mars flyby missions.[8]

Mariner 3 was lost when the launch vehicle's nose fairing failed to jettison. Its sister ship, Mariner 4, launched on November 28, 1964, was the first successful flyby of the planet Mars and gave the first glimpse of Mars at close range.[8]

- Mission: Mars flyby[8]

- Mass: 261 kg (575 lb)

- Mass: 245 kg (540 lb)

- Sensors: ultraviolet photometer, cosmic dust, solar plasma, trapped radiation, cosmic rays, magnetic fields, radio occultation and celestial mechanics

Status: Mariner 5 – Defunct. Now in Heliocentric orbit.

### 1.14.5   Mariners 6 and 7

Main article: Mariner 6 and 7
  Mariners 6 and 7 were identical teammates in a two-

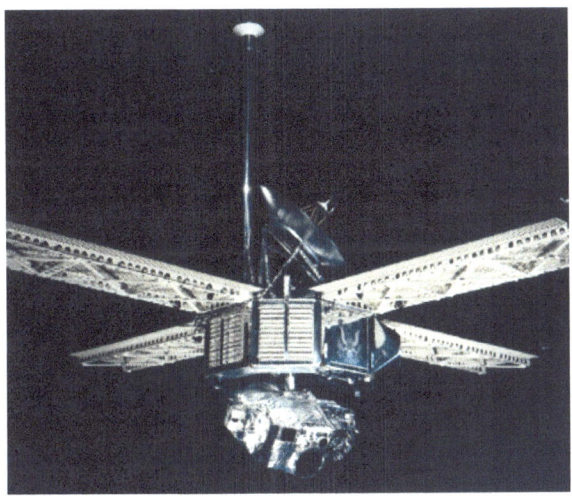

spacecraft mission to Mars. Mariner 6 was launched on February 24, 1969, followed by Mariner 7 on March 21, 1969. They flew over the equator and southern hemisphere of the planet Mars.*[10]

- Mission: Mars flybys

- Mass 413 kg (908 lb)

- Sensors: wide- and narrow-angle cameras with digital tape recorder, infrared spectrometer and radiometer, ultraviolet spectrometer, radio occultation and celestial mechanics.

Status:

- Mariner 6 – Defunct. Now in a Heliocentric orbit.*[10]

- Mariner 7 – Defunct. Now in a Heliocentric orbit.*[10]

### 1.14.6   Mariners 8 and 9

Main articles: Mariner 8 and Mariner 9
Mariner 8 and Mariner 9 were identical sister craft designed

to map the Martian surface simultaneously, but Mariner 8 was lost in a launch vehicle failure. Its identical sister craft, Mariner 9, was launched in May 1971 and became the first artificial satellite of Mars. It entered Martian orbit in November 1971 and began photographing the surface and analyzing the atmosphere with its infrared and ultraviolet instruments.

- Mission: orbit Mars

- Mass 998 kg (2,200 lb)

- Sensors: wide- and narrow-angle cameras with digital tape recorder, infrared spectrometer and radiometer, ultraviolet spectrometer, radio occultation and celestial mechanics

Status:

- Mariner 8 – Destroyed in a launch vehicle failure.

- Mariner 9 – Shut off. In Areocentric (Mars) orbit until at least 2022 when it is projected to fall out of orbit and into the Martian atmosphere.*[11]

### 1.14.7   Mariner 10

Main article: Mariner 10
  The Mariner 10 spacecraft launched on November 3, 1973 and was the first to use a gravity assist trajectory, accelerating as it entered the gravitational influence of Venus, then being flung by the planet's gravity onto a slightly different course to reach Mercury. It was also the first spacecraft to encounter two planets at close range, and for 33 years the only spacecraft to photograph Mercury in closeup.

- Mission: Venus and Mercury flybys

- Mass: 433 kg (952 lb)

- Sensors: twin narrow-angle cameras with digital tape recorder, ultraviolet spectrometer, infrared radiometer, solar plasma, charged particles, magnetic fields, radio occultation and celestial mechanics

Status: Mariner 10 – Defunct. Now in a Heliocentric orbit.

### 1.14.8   Mariner Jupiter-Saturn

Mariner Jupiter-Saturn was approved in 1972 after the cancellation of the Grand Tour program, which proposed visiting all the outer planets with multiple spacecraft. The Mariner Jupiter-Saturn program proposed two Mariner-derived probes that would perform a scaled back mission involving flybys of only the two gas giants, though designers at JPL built the craft with the intention that further encounters past Saturn would be an option. Trajectories were chosen to allow one probe to visit Jupiter and Saturn first, and perform a flyby of Saturn's moon Titan to gather information about the moon's substantial atmosphere. The other probe would arrive at Jupiter and Saturn later, and its trajectory would enable it to continue on to Uranus and Neptune assuming the first probe accomplished all its objectives, or be redirected to perform a Titan flyby if necessary. The program's name was changed to Voyager just before launch in 1977, and after Voyager 1 successfully completed its Titan encounter, Voyager 2 went on to visit the two ice giants.*[3]

### 1.14.9   See also

- Mariner Mark II

- Mariner (crater)

- Pioneer program

- Tom Krimigis

### 1.14.10   References

[1] "Mariner-Venus 1962 Final Project Report" (PDF). NASA Technical Reports Server. Retrieved December 29, 2011.

[2] "Mariner Program". JPL Mission and Spacecraft Library. Retrieved December 28, 2011.

[3] Chapter 11 "Voyager: The Grand Tour of Big Science" (sec. 268.), by Andrew,J. Butrica, found in From Engineering Science To Big Science ISBN 978-0-16-049640-0 edited by Pamela E. Mack, NASA, 1998

[4] Mariner 4, NSSDC Master Catalog

[5] "Untitled" (PDF). NASA Technical Reports Server. Retrieved December 28, 2011.

[6] "Tracking Information Memorandom: Mariner R 1 and 2" (PDF). NASA Technical Reports Server. Retrieved December 29, 2011.

[7] "Mariner R Spacecraft for Missions P-37/P-38" (PDF). NASA Technical Reports Server. Retrieved December 29, 2011.

[8] Pyle, Rod (2012). Destination Mars. Prometheus Books. p. 51. ISBN 978-1-61614-589-7. Mariner 3, dead and still ensnared in its faulty launch shroud, in a large orbit around the sun.

[9] Pyle, Rod (2012). Destination Mars. Prometheus Books. p. 56. ISBN 978-1-61614-589-7. It eventually joined its sibling, Mariner 3, dead ... in a large orbit around the sun.

[10] Pyle, Rod (2012). Destination Mars. Prometheus Books. pp. 61–66. ISBN 978-1-61614-589-7.

[11] NASA - This Month in NASA History: Mariner 9, November 29, 2011 —Vol. 4, Issue 9

## 1.15   Pioneer Venus project

Main articles: Pioneer Venus Orbiter and Pioneer Venus Multiprobe
The **Pioneer Venus** project was part of the Pioneer program consisting of two spacecraft, the Pioneer Venus Orbiter and the Pioneer Venus Multiprobe, launched to Venus in 1978. The program was managed by NASA's Ames Research Center.

The Pioneer Venus Orbiter entered orbit around Venus on December 4, 1978, and performed observations to characterize the atmosphere and surface of Venus. It continued to transmit data until October 1992.

*Cloud structure in the Venusian atmosphere in 1979, revealed by ultraviolet observations by Pioneer Venus Orbiter*

The Pioneer Venus Multiprobe deployed four small probes into the Venusian atmosphere on December 9, 1978. All four probes transmitted data throughout their descent to the surface. One probe survived landing and transmitted data from the surface for over an hour.

### 1.15.1 See also

- 1978 in spaceflight

- Vega program

- Venera program

### 1.15.2 External links

- NASA: Pioneer Venus Project Information

- Pioneer Venus Program Page by NASA's Solar System Exploration

- NSSDC Master Catalog: Spacecraft Pioneer Venus Probe Bus. (Other components of the mission have their own pages at this site too.)

- Several articles in Science (1979), 205, pages 41-121

- Kasprzak, W. T - **The Pioneer Venus Orbiter: 11 years of data.** (May 1, 1990) - NASA

- Art for the mission

## 1.16 Pioneer Venus Multiprobe

The **Pioneer Venus Multiprobe**, also known as **Pioneer Venus 2** or **Pioneer 13** was a spacecraft launched in 1978 to explore Venus as part of NASA's Pioneer program.

### 1.16.1 Spacecraft

*Pioneer Venus Bus with probes attached*

The Pioneer Venus Multiprobe bus was constructed by the Hughes Aircraft Company, built around the HS-507 bus. It was cylindrical in shape, with a diameter of 2.5 metres (8 ft 2 in) and a mass of 290 kilograms (640 lb). Unlike the probes, which did not begin making direct measurements until they had decelerated lower in the atmosphere, the bus returned data on Venus' upper atmosphere.

The bus was targeted to enter the Venusian atmosphere at a shallow entry angle and transmit data until destruction by the heat of atmospheric friction. The objective was to study the structure and composition of the atmosphere down to the surface, the nature and composition of the clouds, the radiation field and energy exchange in the lower atmosphere, and local information on the atmospheric circulation pattern. With no heat shield or parachute, the bus made upper atmospheric measurements with two instruments, an Ion Mass Spectrometer (BIMS) and a Neutral Mass Spectrometer (BNMS), down to an altitude of about

110 km before disintegrating on December 9, 1978.

## 1.16.2  Probes

The spacecraft carried one large and three small atmospheric probes, designed to collect data as they descended into the atmosphere of Venus. The probes did not carry photographic instruments, and were not designed to survive landing - the smaller probes were not equipped with parachutes, and the larger probe's parachute was expected to detach as it neared the ground. All four probes continued transmitting data until impact; however, one survived and continued to transmit data from the surface.

**Large probe**

*Pioneer Venus Large Probe opens its parachute*

The Large probe carried seven experiments, contained within a sealed spherical pressure vessel. The science experiments were:

- a neutral mass spectrometer to measure the atmospheric composition

- a gas chromatograph to measure the atmospheric composition

- a solar flux radiometer to measure solar flux penetration in the atmosphere

- an infrared radiometer to measure distribution of infrared radiation

- a cloud particle size spectrometer to measure particle size and shape

- a nephelometer to search for cloud particles

- temperature, pressure, and acceleration sensors

This pressure vessel was encased in a nose cone and aft protective cover. After deceleration from initial atmospheric entry at about 11.5 kilometres per second (7.1 mi/s) near the equator on the night side of Venus, a parachute was deployed at 47 km altitude. The large probe was about 150 centimetres (59 in) in diameter and the pressure vessel itself was 73.2 centimeters (28.8 in) in diameter.

**Small probes**

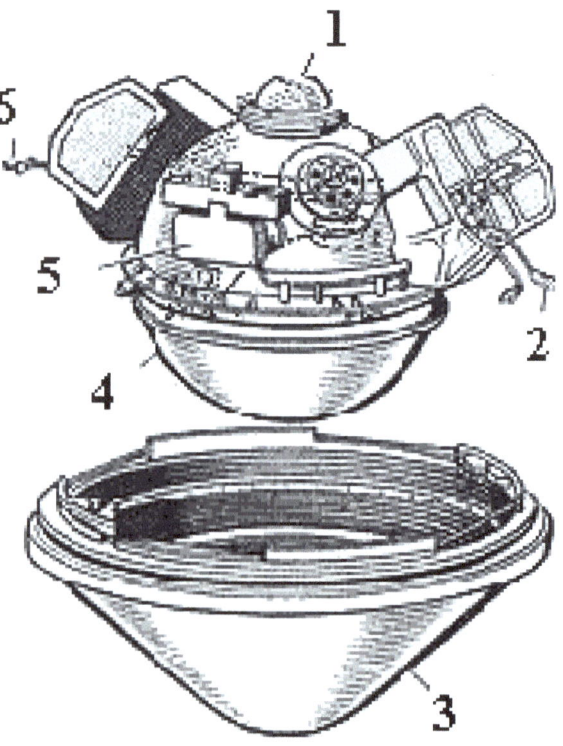

*A small probe (1-antenna, 2-temperature sensor, 3-frontal protection, 4-hermetic container, 5-nephelometer, 6-radiometer)*

Three identical small probes, around 0.8 metres (2 ft 7 in) in diameter, were deployed. These probes consisted of spherical pressure vessels surrounded by an aeroshell, but unlike the large probe, they had no parachutes and the aeroshells did not separate from the probes.

Each probe carried a nephelometer and temperature, pressure, and acceleration sensors, as well as a net flux radiometer experiment to map the distribution of sources and sinks of radiative energy in the atmosphere. The radio signals from all four probes were also used to characterize the winds, turbulence, and propagation in the atmosphere.

The small probes were each targeted at different parts of the planet and were named accordingly.

- The **North probe** entered the atmosphere at about 60 degrees north latitude on the day side.

- The **Night probe** entered on the night side.

- The **Day probe** entered well into the day side, and was the only one of the four probes which continued to send radio signals back after impact, for over an hour.

### 1.16.3 Launch

The Pioneer Venus Multiprobe was launched by an Atlas SLV-3D Centaur-D1AR rocket, which flew from Launch Complex 36A at the Cape Canaveral Air Force Station. The launch occurred at 07:33 on August 8, 1978, and deployed the Multiprobe into heliocentric orbit for its coast to Venus.

### 1.16.4 Arrival at Venus

Prior to the Multiprobe reaching Venus, the four probes were deployed from the main bus. The large probe was released on November 16, 1978, and the three small probes on November 20.

All four probes and the bus reached Venus on December 9, 1978. The large probe was the first to enter the atmosphere, at 18:45:32 UTC, followed over the next 11 minutes by the other three probes. The bus entered the atmosphere at 20:21:52 UTC, and returned its last signal at 20:22:55 from an altitude of 110 kilometres (68 mi).

The four probes transmitted data until they impacted the surface of Venus. The Day Probe survived impact, returning data from the surface for over an hour.

### 1.16.5 See also

- Timeline of artificial satellites and space probes

### 1.16.6 External links

- NASA: Pioneer Venus Project Information

- Pioneer Venus Program Page by NASA's Solar System Exploration

- NSSDC Master Catalog: Spacecraft Pioneer Venus Probe Bus. (Other components of the mission have their own pages at this site too.)

- Several articles in Science (1979), 205, pages 41-121

## 1.17 Pioneer Venus Orbiter

The **Pioneer Venus Orbiter**, also known as **Pioneer Venus 1** or **Pioneer 12**, was a mission to Venus conducted as part of the Pioneer program. Launched in May 1978 atop an Atlas-Centaur rocket, the spacecraft was inserted into an elliptical orbit around Venus on December 4, 1978. It returned data on Venus until October 1992.[*][1][*][2]

### 1.17.1 Launch and arrival at Venus

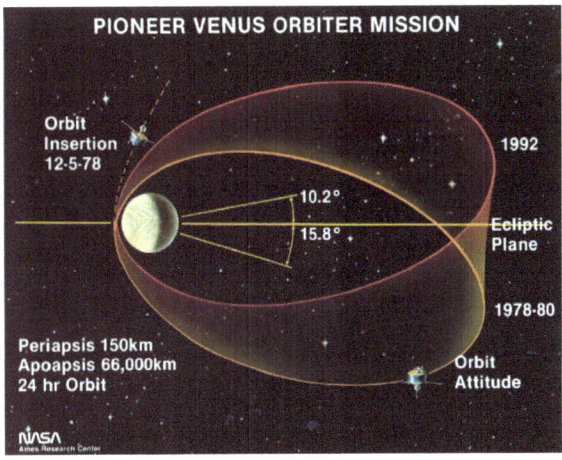

*Orbit attitude of Pioneer Venus 1 between 1978 - 1980 and 1992*

The Pioneer Venus Orbiter was launched by an Atlas SLV-3D Centaur-D1AR rocket, which flew from Launch Complex 36A at the Cape Canaveral Air Force Station. The launch occurred at 13:13:00 on May 20, 1978, and deployed the Orbiter into heliocentric orbit for its coast to Venus. Venus orbit insertion occurred on December 4, 1978.

### 1.17.2 Spacecraft

Manufactured by Hughes Aircraft Company, the Pioneer Venus Orbiter was based on the HS-507 bus.[*][3] The spacecraft was a flat cylinder, 2.5 meters (8.2 ft) in diameter and 1.2 meters (3.9 ft) long. All instruments and spacecraft subsystems were mounted on the forward end of the cylinder,

*Pioneer Venus 1 at KSC*

### 1.17.3 Experiments

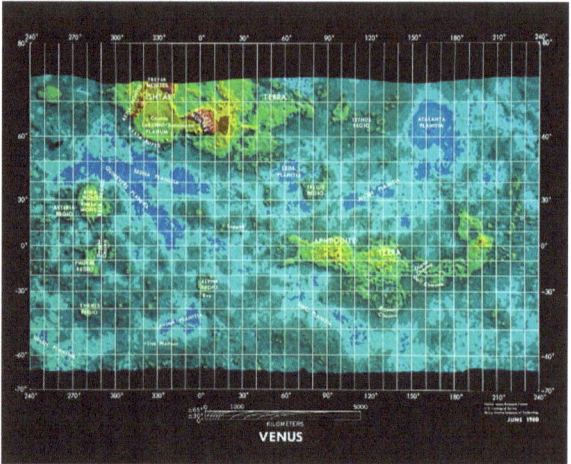

*A map of Venus produced from Pioneer data*

The Pioneer Venus Orbiter carried 17 experiments (with a total mass of 45 kg):

*An image of Venus in ultraviolet light by the Pioneer Venus Orbiter*

except the magnetometer, which was at the end of a 4.7 meters (15 ft) boom. A solar array extended around the circumference of the cylinder. A 1.09 metres (3 ft 7 in) despun dish antenna provided S and X band communication with Earth. A Star-24 solid rocket motor was integrated into the spacecraft to provide the thrust to enter orbit around Venus.[*][3]

From Venus orbit insertion to July 1980, periapsis was held between 142 and 253 kilometres (88 and 157 mi) (at 17 degrees north latitude) to facilitate radar and ionospheric measurements. The spacecraft was in a 24-hour orbit with an apoapsis of 66,900 kilometers (41,600 mi). Thereafter, the periapsis was allowed to rise to a maximum of 2,290 kilometres (1,420 mi) and then fall, to conserve fuel.

In 1991, the Radar Mapper was reactivated to investigate previously inaccessible southern portions of the planet, in conjunction with the recently arrived Magellan spacecraft. In May 1992, Pioneer Venus began the final phase of its mission, in which the periapsis was held between 150 and 250 kilometres (93 and 155 mi), until the spacecraft's propellant was exhausted, after which the orbit decayed naturally. The spacecraft continued to return data until 8 October 1992, with the last signals being received at 19:22 UTC.[*][2] The Pioneer Venus Orbiter disintegrated upon entering the atmosphere of Venus on October 22, 1992.[*][1]

- a **cloud photo-polarimeter (OCPP)** to measure the vertical distribution of the clouds, similar to Pioneer 10 and Pioneer 11 imaging photo-polarimeter (IPP)

- a **surface radar mapper (ORAD)** to determine topography and surface characteristics. Observations could only be conducted when the probe was closer than 4700 km over the planet. A 20 Watt S-band signal (1.757 gigahertz) was sent to the surface that reflected it, with the probe analyzing the echo. Resolution at periapsis was 23 x 7 km.

- an **infrared radiometer (OIR)** to measure IR emissions from Venus' atmosphere

- an airglow **ultraviolet spectrometer (OUVS)** to measure scattered and emitted UV light

- a **neutral mass spectrometer (ONMS)** to determine the composition of the upper atmosphere

- a solar wind **plasma analyzer (OPA)** to measure properties of the solar wind

- a **magnetometer (OMAG)** to characterize the magnetic field at Venus

- an **electric field detector (OEFD)** to study the solar wind and its interactions

- an **electron temperature (OETP)** to study the thermal properties of the ionosphere

- an **ion mass spectrometer (OIMS)** to characterize the ionospheric ion population

- a **charged particle retarding potential analyzer (ORPA)** to study ionospheric particles

- two radio science experiments to determine the gravity field of Venus

- a radio occultation experiment to characterize the atmosphere

- an atmospheric drag experiment to study the upper atmosphere

- a radio science atmospheric and solar wind turbulence experiment

- a **gamma ray burst (OGBD)** detector to record gamma ray burst events

The spacecraft conducted radar altimetry observations allowing the first global topographic map of the Venusian surface to be constructed.

### 1.17.4 Observations of Halley's Comet

From its orbit of Venus, the Pioneer Venus Orbiter was able to observe Halley's Comet when it was unobservable from Earth due to its proximity to the sun during February 1986. UV spectrometer observations monitored the loss of water from the comet's nucleus at perihelion on February 9.[*][4]

### 1.17.5 See also

- Timeline of artificial satellites and space probes

### 1.17.6 References

[1] McDowell, Jonathan. "Satellite Catalog". *Jonathan's Space Page*. Archived from the original on 2003-10-11.

[2] "Pioneer Venus 1". *Solar System Exploration*. NASA. Archived from the original on 2006-10-04.

[3] Krebs, Gunter. "Pioneer 12 (Pioneer Venus Orbiter, PVO)". *Gunter's Space Page*. Archived from the original on 2005-01-12.

[4] Russell, C.T.; Luhmann, J.G.; Scarf, F.L. (1985). "Pioneer Venus Observations during Comet Halley's Inferior Conjunction" (PDF). University of California, Los Angeles. Archived (PDF) from the original on 2009-02-27.

### 1.17.7 External links

- NASA: Pioneer Venus Project Information

- Pioneer Venus Program Page by NASA's Solar System Exploration

- Several articles in Science (1979), 205, pages 41-121

- Kasprzak, W. T - **The Pioneer Venus Orbiter: 11 years of data.** (May 1, 1990) - NASA

## 1.18 TMK

For other uses, see TMK (disambiguation).

**TMK** (Russian: Тяжелый Межпланетный Корабль -

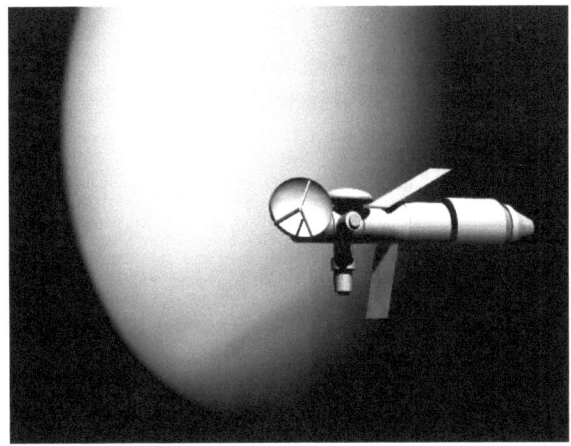

*Artist's depiction of TMK-MAVR on a Venus flyby*

*Tyazhely Mezhplanetny Korabl* for *Heavy Interplanetary Spacecraft*) was the designation of a Soviet space exploration project to send a manned flight to Mars and Venus (TMK-MAVR design) without landing.

The TMK-1 spacecraft was due to be launched in 1971 and make a three-year-long flight including a Mars flyby, at which time probes would have been dropped. Expanded project variations, such as the TMK-E, Mavr or KK, including a Venus flyby, electric propulsion or a manned Mars landing were also proposed.

The TMK project was planned as an answer from the Soviet Union to the United States' manned Moon landings. A previous *Martian Piloted Complex* mission was proposed in 1956. The project was never completed because the required N1 rocket never flew successfully.

## 1.18.1   TMK-1

The first flight to Mars of the TMK-1 was planned to begin on June 8, 1971.

The 75 metric ton TMK-1 spacecraft would take a crew of three on a Mars flyby mission. After a 10½ month flight the crew would race past Mars, dropping remote-controlled landers, and then be flung into an Earth-return trajectory. Earth return would happen on July 10, 1974, after a voyage of three years, one month, and two days.

Spacecraft configuration:

- A *habitation* or *pilot compartment*, with an internal volume of 25 cubic meters

- A *work* or *equipment section*, including the hatch for extra-vehicular activities and a solar storm shelter should solar flares increase radiation to dangerous levels. Total volume of the section would be 25 cubic meters.

- A *biological systems compartment*, with the SOZh closed-cycle environmental control system, with a total volume of 75 cubic meters

- A *aggregate section*, with the Mars probe capsules, the KDU midcourse correction engine, the SOZh solar concentrator and solar panels, and radio antennas

- The *SA* crew Earth reentry capsule, about 4 m in diameter

## 1.18.2   TMK-E

This variation was proposed in 1960, and consisted of a complete Mars landing expedition to be assembled in Earth

orbit using several N1 launches. The spacecraft would be powered by nuclear electric engines and five landers would deliver a nuclear-powered *Mars Train* on the surface for a one-year mission.

The TMK-E would be capable of a three-year flight to Mars and return, of which one year was powered flight. It would measure 175 m in length and house a crew of six. Six landing craft were included, two for the crew and four for the *Mars Train* vehicles.

## 1.18.3   Mavr (MArs - VeneRa)

A variation of the TMK mission planning involved a flyby of Venus on the return voyage, and was given the code name "Mavr" (*MArs - VeneRa*), meaning Mars - Venus.

## 1.18.4   KK - Space Complex for Delivering a Piloted Expedition to Mars

In 1966, a final version of the TMK studies was known as *KK - Space Complex for Delivering a Piloted Expedition to Mars*. Nuclear electric propulsion was to be used for the 630 day mission. The craft structure consisted of:

- *EK - Expeditionary spacecraft*: command center for piloting in interplanetary space

- *OK - Orbital Complex*: living and work compartments and the life support systems

- *SA - The Landing Module, AV - The Ascent Module* and *RV - The Ascent Rocket stage*

- *PS - The Planetary Station*: used by the expedition on the Martian surface for life support and scientific research

The launch was planned for 1980, with a crew of three cosmonauts. Mars stay duration would be 30 days.

Mission data:

- Total Payload Required in Low Earth Orbit-metric tons: 150

- Total Propellant Required-metric tons: 24

- Number of Launches Required to Assemble Payload in Low Earth Orbit: 2

- Launch Vehicle: N1

### 1.18.5  See also

- Manned mission to Mars

- Manned Venus Flyby

### 1.18.6  External links

- Anatoly Zak's TMK page

- TMK page on astronautix

- TMK-MAVR page on astronautix

## 1.19  Tyazhely Sputnik

**Tyazhely Sputnik**, (Russian: Тяжелый Спутник meaning *Heavy Satellite*), also known by its development name as **Venera 1VA No.1**,[*][1] and in the West as **Sputnik 7**, was a Soviet spacecraft, which was intended to be the first spacecraft to explore Venus. Due to a problem with its upper stage it failed to leave low Earth orbit. In order to avoid acknowledging the failure, the Soviet government instead announced that the entire spacecraft, including the upper stage, was a test of a "Heavy Satellite" which would serve as a launch platform for future missions. This resulted in the upper stage being considered a separate spacecraft, from which the probe was "launched", on several subsequent missions.[*][2]

Tyazhely Sputnik was launched at 01:18:03 UTC on 4 February 1961, atop a Molniya 8K78 carrier rocket flying from Site 1/5 at the Baikonur Cosmodrome.[*][3] When the upper stage ignited, cavitation in the liquid oxygen flowing through the oxidiser pump caused the pump to fail, resulting in an engine failure eight-tenths of a second after ignition.[*][4] It reentered the atmosphere over Siberia on 26 February 1961.[*][5]

According to the memoirs of Boris Chertok, "...A pendant shaped like a small globe with the continents etched on it was placed on the 1VA. Inside this small sphere was a medal depicting the Earth-to-Venus flight path. On the other side of the medal was the emblem of the Soviet Union. The pendant was placed in a spherical capsule with thermal shielding to protect it during entry into Venus' atmosphere at reentry velocity." [*][6] In what he refers to as a "Strange but True [incident]...in the history of cosmonautics," while the spacecraft was originally thought to have re-entered over the Pacific ocean, it was subsequently (in 1963) found to have re-entered over Siberia, when this medal made its way back to Chertok by way of his boss, Chief Rocket Designer Sergei Korolev. He relates that, "while swimming in a river

—a tributary of the Biryusa River in eastern Siberia —a local boy hurt his foot on some sort of piece of iron. When he retrieved it from the water, rather than throw it into deeper water, he brought it home and showed it to his father. The boy's father, curious as to what the dented metal sphere contained, opened it up and discovered this medal inside ···The boy's father brought his find to the police. The local police delivered the remains of the pendant to the regional department of the KGB, which in turn forwarded it to Moscow. In Moscow the appropriate KGB directorate··· after notifying Keldysh as president of the Academy of Sciences," delivered the pendant to Korolev. "Thus, [Chertok] was awarded the medal that had been certified for the flight to Venus by the protocol that [he and Korolev] signed in January 1961. After the launch we were all certain that the Tyazhelyy sputnik and the pendant had sunk in the ocean. Now it turned out that it had burned up over Siberia. The pendant had been designed to withstand Venus' atmosphere and therefore it reached the Earth's surface."

The sister probe, Venera 1, successfully launched and was injected into a heliocentric orbit toward Venus one week later, although telemetry on the mission failed a week into flight.

### 1.19.1  See also

- Venera

### 1.19.2  References

[1] Zak, Anatoly. "Russia's unmanned missions to Venus". RussianSpaecWeb. Retrieved 10 January 2011.

[2] Wade, Mark. "Venera 1VA". Encyclopedia Astronautica. Retrieved 28 July 2010.

[3] McDowell, Jonathan. "Launch Log". Jonathan's Space Page. Retrieved 28 July 2010.

[4] Wade, Mark. "Soyuz". Encyclopedia Astronautica. Retrieved 28 July 2010.

[5] "Sputnik 7". NSSDC (NASA Goddard Spaceflight Center). Retrieved 28 July 2010.

[6] Chertok, Boris. *Creating a Rocket Industry* (PDf). pp. 578 and 585–586.

## 1.20  Vega 1

**Vega 1** (along with its twin Vega 2) is a Soviet space probe part of the Vega program. The spacecraft was a development of the earlier *Venera* craft. They were designed by Babakin Space Centre and constructed as **5VK** by Lavochkin at Khimki.

The craft was powered by twin large solar panels and instruments included an antenna dish, cameras, spectrometer, infrared sounder, magnetometers (MISCHA), and plasma probes. The 4,920 kg craft was launched by a Proton 8K82K rocket from Baikonur Cosmodrome, Tyuratam, Kazakh SSR. Both Vega 1 and 2 were three-axis stabilized spacecraft. The spacecraft were equipped with a dual bumper shield for dust protection from Halley's comet.

## 1.20.1   Venus mission

The descent module arrived at Venus on 11 June 1985, two days after being released from the Vega 1 flyby probe. The module, a 1500 kg, 240 cm diameter sphere, contained a surface lander and a balloon explorer. The flyby probe performed a gravitational assist maneuver using Venus, and continued its mission to intercept the comet.[1]

### Descent craft

The surface lander was identical to that of Vega 2 as well as the previous six *Venera* missions. The objective of the probe was the study of the atmosphere and the exposed surface of the planet. The scientific payload included a UV spectrometer, temperature and pressure sensors, a water concentration meter, a gas-phase chromatograph, an X-ray spectrometer, a mass spectrometer, and a surface sampling device. Several of these scientific tools (the UV spectrometer, the mass spectrograph, and the devices to measure pressure and temperature) were developed in collaboration with French scientists.[1]

The lander successfully touched down at 7°12′N 177°48′E / 7.2°N 177.8°E in the Mermaid Plain north of Aphrodite Terra. Due to excessive turbulence, some surface experiments were inadvertently activated 20 km above the surface. Only the mass spectrometer was able to return data.[2]

### Balloon

The Vega 1 Lander/Balloon capsule entered the Venus atmosphere (125 km altitude) at 2:06:10 UT (Earth received time; Moscow time 5:06:10 a.m.) on 11 June 1985 at roughly 11 km/s. At approximately 2:06:25 UT the parachute attached to the landing craft cap opened at an altitude of 64 km. The cap and parachute were released 15 seconds later at 63 km altitude. The balloon package was pulled out of its compartment by parachute 40 seconds later

at 61 km altitude, at 8.1 degrees N, 176.9 degrees east. A second parachute opened at an altitude of 55 km, 200 seconds after entry, extracting the furled balloon. The balloon was inflated 100 seconds later at 54 km and the parachute and inflation system were jettisoned. The ballast was jettisoned when the balloon reached roughly 50 km and the balloon floated back to a stable height between 53 and 54 km some 15 to 25 minutes after entry. The mean stable height was 53.6 km, with a pressure of 535 mbar and a temperature of 300-310 K in the middle, most active layer of the Venus three-tiered cloud system. The balloon drifted westward in the zonal wind flow with an average speed of about 69 m/s (248 km/hr) at nearly constant latitude. The probe crossed the terminator from night to day at 12:20 UT on 12 June after traversing 8500 km. The probe continued to operate in the daytime until the final transmission was received at 00:38 UT on 13 June from 8.1 N, 68.8 E after a total traverse distance of 11,600 km or about 30% of the circumference of the planet. It is not known how much farther the balloon travelled after the final communication.[2]

## 1.20.2   Halley mission

After their encounters, the Vegas' motherships were redirected by Venus' gravity to intercept Halley's Comet.

Images started to be returned on March 4, 1986, and were used to help pinpoint Giotto's close flyby of the comet. The early images from Vega showed two bright areas on the comet, which were initially interpreted as a double nucleus. The bright areas would later turn out to be two jets emitting from the comet. The images also showed the nucleus to be dark, and the infrared spectrometer readings measured a nucleus temperature of 300 K to 400 K, much warmer than expected for an ice body. The conclusion was that the comet had a thin layer on its surface covering an icy body.

Vega 1 made its closest approach on March 6 at around 8,889 kilometers (at 07:20:06 UT) of the nucleus. It took more than 500 pictures via different filters as it flew through the gas cloud around the coma. Although the spacecraft was battered by dust, none of the instruments were disabled during the encounter.

The data intensive examination of the comet covered only the three hours around closest approach. They were intended to measure the physical parameters of the nucleus, such as dimensions, shape, temperature and surface properties, as well as to study the structure and dynamics of the coma, the gas composition close to the nucleus, the dust particles' composition and mass distribution as functions of distance to the nucleus and the cometary-solar wind interaction.

The Vega images showed the nucleus to be about 14 km

long with a rotation period of about 53 hours. The dust mass spectrometer detected material similar to the composition of carbonaceous chondrites meteorites and also detected clathrate ice.

After subsequent imaging sessions on 7 and 8 March 1986, Vega 1 headed out to deep space. In total Vega 1 and Vega 2 returned about 1500 images of Comet Halley. Vega 1 ran out of attitude control propellant on 30 January 1987, and contact with Vega 2 continued until 24 March 1987.

Vega 1 and Vega 2 are currently in heliocentric orbits.

### 1.20.3 See also

- Venera program

- Vega program

### 1.20.4 External links

- Vega 1 Measuring Mission Profile by NASA's Solar System Exploration

- Vega mission images from the Space Research Institute (IKI)

- Raw data from Vega 1 and Vega 2 on board instruments

- Soviet Exploration of Venus

### 1.20.5 References

[1] NASA - NSSDC - Spacecraft - Details

[2] NASA Database - Solar System Exploration; Missions; By Target; Venus; Past; Vega 1

## 1.21 Vega 2

**Vega 2** (along with Vega 1) is a Soviet space probe part of the Vega program. The spacecraft was a development of the earlier *Venera* craft. They were designed by Babakin Space Centre and constructed as **5VK** by Lavochkin at Khimki. The craft was powered by twin large solar panels and instruments included an antenna dish, cameras, spectrometer, infrared sounder, magnetometers (MISCHA), and plasma probes. The 4,920 kg craft was launched by a Proton 8K82K rocket from Baikonur Cosmodrome, Tyuratam, Kazakh SSR. Both Vega 1 and 2 were three-axis stabilized spacecraft. The spacecraft were equipped with a dual bumper shield for dust protection from Halley's Comet.

### 1.21.1 Venus mission

The descent module arrived at Venus on 15 June 1985, two days after being released from the Vega 2 flyby probe. The module, a 1500 kg, 240 cm diameter sphere, contained a surface lander and a balloon explorer. The flyby probe performed a gravitational assist maneuver using Venus, and continued its mission to intercept the comet.[1]

**Lander**

The surface lander was identical to that of Vega 1 as well as the previous six *Venera* missions. The objective of the probe was the study of the atmosphere and the exposed surface of the planet. The scientific payload included a UV spectrometer, temperature and pressure sensors, a water concentration meter, a gas-phase chromatograph, an X-ray spectrometer, a mass spectrometer, and a surface sampling device. Several of these scientific tools (the UV spectrometer, the mass spectrograph, and the devices to measure pressure and temperature) were developed in collaboration with French scientists.[1]

The Vega 2 lander touched down at 03:00:50 UT on 15 June 1985 at around 7°08′S 177°40′E / 7.14°S 177.67°E, in the northern region of Aphrodite Terra. The altitude of the touchdown site was 0.1 km above the planetary mean radius. The measured pressure at the landing site was 91 atm and the temperature was 736 K. The surface sample was found to be an anorthosite-troctolite rock, rarely found on Earth, but present in the lunar highlands, leading to the conclusion that the area was probably the oldest explored by any Venera vehicle. It transmitted data from the surface for 56 minutes.[2]

**Balloon**

The Vega 2 Lander/Balloon capsule entered the Venus atmosphere (125 km altitude) at 02:06:04 UT (Earth received time; Moscow time 05:06:04) on 15 June 1985 at roughly 11 km/s. At approximately 2:06:19 UT the parachute attached to the landing craft cap opened at an altitude of 64 km. The cap and parachute were released 15 seconds later at 63 km altitude. The balloon package was pulled out of its compartment by parachute 40 seconds later at 61 km altitude, at 7.45 degrees S, 179.8 degrees east. A second parachute opened at an altitude of 55 km, 200 seconds after entry, extracting the furled balloon. The balloon was inflated 100 seconds later at 54 km and the parachute and inflation system were jettisoned. The ballast was jettisoned when the balloon reached roughly 50 km and the balloon floated back to a stable height between 53 and 54 km some 15 to 25 minutes after entry. The mean stable height was

53.6 km, with a pressure of 535 mbar and a temperature of 308-316 K in the middle, most active layer of the Venus three-tiered cloud system. The balloon drifted westward in the zonal wind flow with an average speed of about 66 m/s at nearly constant latitude. The probe crossed the terminator from night to day at 9:10 UT on 16 June after traversing 7400 km. The probe continued to operate in the daytime until the final transmission was received at 00:38 UT on 17 June from 7.5 S, 76.3 E after a total traverse distance of 11,100 km. It is not known how much further the balloon traveled after the final communication.*[2]

### 1.21.2   Halley mission

After their encounters, the Vegas' motherships were redirected by Venus' gravity to intercept Halley's Comet.

The spacecraft initiated its encounter on March 7, 1986 by taking 100 photos of the comet from a distance of 14 million kilometers.

Vega 2 made its closest approach at 07:20 UT on March 9, 1986 at 8,030 km. The data intensive examination of the comet covered only the three hours around closest approach. They were intended to measure the physical parameters of the nucleus, such as dimensions, shape, temperature and surface properties, as well as to study the structure and dynamics of the coma, the gas composition close to the nucleus, the dust particles' composition and mass distribution as functions of distance to the nucleus and the cometary-solar wind interaction.

During the encounter, Vega 2 took 700 images of the comet, with better resolution than those from the twin Vega 1, partly due to the presence of less dust outside the coma at the time. Yet Vega 2 recorded an 80% power loss during the encounter as compared to Vega 1's 40%.

After further imaging sessions on 10 and 11 March 1986, Vega 2 finished its primary mission. Vega 1 ran out of attitude control propellant on 30 January 1987, and contact with Vega 2 continued until 24 March 1987.

Vega 2 is currently in heliocentric orbit.

### 1.21.3   See also

- Venera program

- Vega program

### 1.21.4   References

[1]  NASA - NSSDC - Spacecraft - Details

[2]  NASA Database - Solar System Exploration; Missions; By Target; Venus; Past; Vega 2

### 1.21.5   External links

- Vega 2 Measuring Mission Profile by NASA's Solar System Exploration

- Vega mission images from the Space Research Institute (IKI)

- Raw data from Vega 1 and Vega 2 on board instruments

- Soviet Exploration of Venus

## 1.22   Venera

*For other uses, see Venera (disambiguation).*

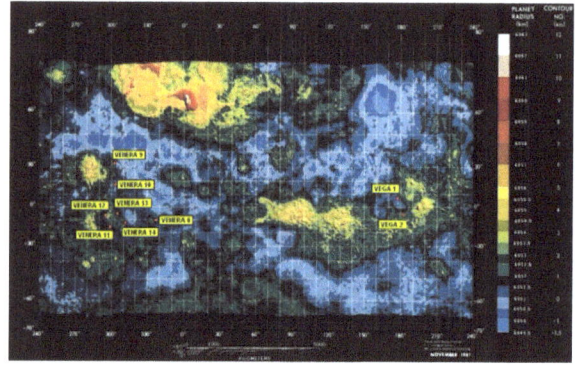

*Location of Soviet Venus landers*

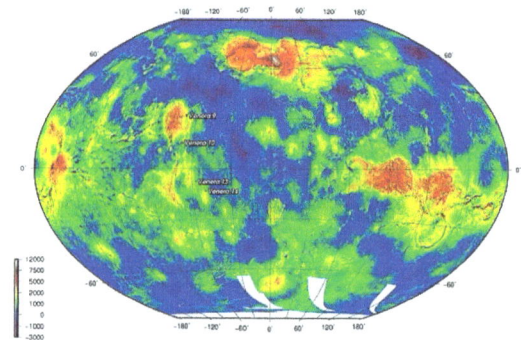

*Position of Venera landing sites returning images from the surface*

The **Venera** (Russian: Венера, pronounced [vʲɪˈnʲɛrə]) series space probes were developed by the Soviet Union between 1961 and 1984 to gather data from Venus, Venera being the Russian name for Venus. As with some of the Soviet Union's other planetary probes, the later versions were launched in pairs with a second vehicle being launched soon after the first of the pair.

Ten probes from the Venera series successfully landed on Venus and transmitted data from the surface of Venus, including the two Vega program and Venera-Halley probes. In addition, thirteen Venera probes successfully transmitted data from the atmosphere of Venus.

Among the other results, probes of the series became the first human-made devices to enter the atmosphere of another planet (Venera 4 on October 18, 1967), to make a soft landing on another planet (Venera 7 on December 15, 1970), to return images from the planetary surface (Venera 9 on June 8, 1975), and to perform high-resolution radar mapping studies of Venus (Venera 15 on June 2, 1983). The entire series could be considered highly successful. The surface conditions on Venus are extreme, therefore the probes only survived on the surface for a duration of 23 minutes (initial probes) up to about two hours (final probes).

## 1.22.1 The Venera probes

Main articles: Venera 1 and Venera 2

The first Soviet attempt at a flyby probe to Venus was launched on February 4, 1961, but failed to leave Earth orbit. In keeping with the (then) Soviet policy of not announcing details on failed missions, the launch was announced under the name Tyazhely Sputnik ( "Heavy Satellite" ). It is also known as Venera 1VA.[*][1]

Venera 1 and Venera 2 were intended as fly-by probes to fly past Venus without entering orbit. Venera 1 was launched on February 12, 1961. Telemetry on the probe failed seven days after launch. It is believed to have passed within 100,000 km of Venus and entered heliocentric orbit. Venera 2 launched on November 12, 1965, but also suffered a telemetry failure after leaving Earth orbit.

Several other failed attempts at Venus flyby probes were launched by the Soviet Union in the early 1960s,[*][2][*][3] but were not announced as planetary missions at the time, and hence did not officially receive the "Venera" designation.

### Venera 3 to 6

Main articles: Venera 3, Venera 4, Venera 5 and Venera 6

The Venera 3 to 6 probes were similar. Weighing approximately one ton, and launched by the Molniya-type booster rocket, they included a cruise "bus" and a spherical atmospheric entry probe. The probes were optimised for atmospheric measurements, but not equipped with any special landing apparatus. Although it was hoped they would reach the surface still functioning, the first probes failed almost immediately, thereby disabling data transmission to Earth.

Venera 3 became the first human-made object to impact another planet's surface as it crash-landed on March 1, 1966. However, as the spacecraft's dataprobes had failed upon atmospheric penetration, no data from within the Venusian boundary were retrieved from the mission.

On 18 October 1967, Venera 4 became the first spacecraft to measure the atmosphere of another planet. While the Soviet Union initially claimed the craft reached the surface intact, re-analysis including atmospheric occultation data from the American Mariner 5 spacecraft that flew by Venus the day after its arrival demonstrated that Venus's surface pressure was 75-100 atmospheres, much higher than Venera 4's 25 atm hull strength, and the claim was retracted.

Realizing the ships would be crushed before reaching the surface, the Soviets launched Venera 5 and Venera 6 as atmospheric probes. Designed to jettison nearly half their payload prior to entering the planet's atmosphere, these craft recorded 53 and 51 minutes of data, respectively, while slowly descending by parachute before their batteries failed.

### Venera 7

Main article: Venera 7

The Venera 7 probe was the first one designed to survive Venus surface conditions and to make a soft landing. Massively overbuilt to ensure survival, it had few experiments on board, and scientific output from the mission was further limited due to an internal switchboard failure which stuck in the "transmit temperature" position. Still, the control scientists succeeded in extrapolating the pressure (90 atm) from the temperature data (465 °C (869 °F)), which resulted from the first direct surface measurements. The Doppler measurements of the Venera 4 to 7 probes were the first evidence of the existence of high-speed zonal winds (up to 100 metres per second (330 ft/s) or 362 kilometres per hour (225 mph)) in the Venusian atmosphere (super rotation).

Venera 7's parachute failed shortly before landing very close to the surface. It impacted at 17 metres per second (56 ft/s) and toppled over, but survived. Due to the resultant antenna misalignment, the radio signal was very weak, but was detected (with temperature telemetry) for 23 more minutes before its batteries expired. Thus, it became, on 15 December 1970, the first human-made probe to transmit data from the surface of Venus.

## Venera 8

Main article: Venera 8

Venera 8 was equipped with an extended set of scientific instruments for studying the surface (gamma-spectrometer etc.). The cruise bus of Venera 7 and 8 was similar to that of earlier ones, with the design ascending to the Zond 3 mission. The lander transmitted data during the descent and landed in sunlight. It measured the light level but had no camera. It continued to send back data for almost an hour.

## Venera 9 to 12

Main articles: Venera 9, Venera 10, Venera 11 and Venera 12
The Venera 9 to 12 probes were of a different design. They

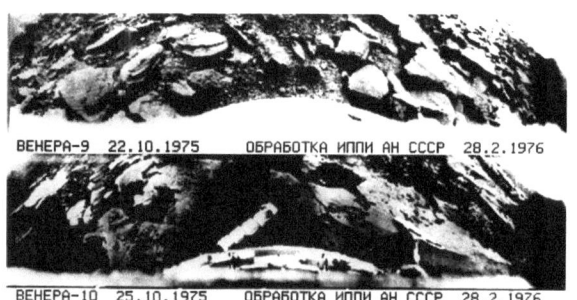

*Surface of Venus imaged by Venera 9 (top) and Venera 10 (bottom)*

weighed approximately five tons and were launched by the powerful Proton booster. They included a transfer and relay bus that had engines to brake into Venus orbit (Venera 9 and 10, 15 and 16) and to serve as receiver and relay for the entry probe's transmissions. The entry probe was attached to the top of the bus in a spherical heat shield. The probes were optimized for surface operations with an unusual looking design that included a spherical compartment to protect the electronics from atmospheric pressure and heat for as long as possible. Beneath this was a shock absorbing "crush ring" for landing. Above the pressure sphere was a cylindrical antenna structure and a wide dish shaped structure that resembled an antenna but was actually an aerobrake. They were designed to operate on the surface for a minimum of 30 minutes. Instruments varied on different missions, but included cameras and atmospheric and soil analysis equipment. All four landers had problems with some or all of their camera lens caps not releasing.

The Venera 9 lander operated for at least 53 minutes and

took pictures with one of two cameras; the other lens cap did not release.

The Venera 10 lander operated for at least 65 minutes and took pictures with one of two cameras; the other lens cap did not release.

The Venera 11 lander operated for at least 95 minutes but neither camera's lens cap released.

The Venera 12 lander operated for at least 110 minutes but neither camera's lens cap released.

## Venera 13 and 14

Main articles: Venera 13 and Venera 14
The descent craft/lander contained most of the instrumen-

*Surface of Venus imaged by Venera 14*

tation and electronics, and was topped by an antenna. The design was similar to the earlier Venera 9–12 landers. They carried instruments to take scientific measurements of the ground and atmosphere once landed, including cameras, a microphone, a drill and surface sampler, and a seismometer. They also had instruments to record electric discharges during its descent phase through the Venusian atmosphere.

The two descent craft landed about 950 kilometres (590 mi) apart, just east of the eastern extension of an elevated region known as Phoebe Regio. The Venera 13 lander survived for 127 minutes, and the Venera 14 lander for 57 minutes, where the planned design life was only 32 minutes. The Venera 14 craft had the misfortune of ejecting the camera lens cap directly under the surface compressibility tester arm, and returned information for the compressibility of the lens cap rather than the surface. The descent vehicles transmitted data to the buses, which acted as data relays as they flew by Venus.

## Veneras 15 and 16

Main articles: Venera 15 and Venera 16
Venera 15 and 16 were similar to previous probes, but replaced the entry probes with surface imaging radar equip-

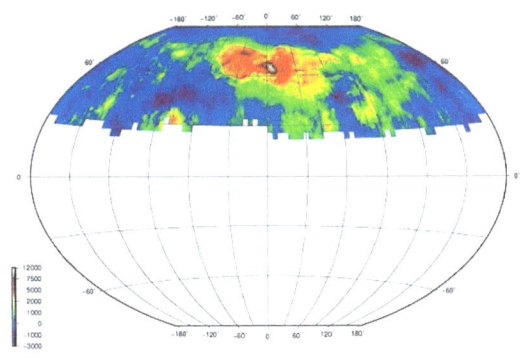

*Radar topography obtained by Venera 15/16*

ment. Radar imaging was necessary to penetrate the dense cloud of Venus.

**Vega probes**

Main article: Vega program

The Vega probes to Venus and comet Halley launched in 1985 also used this basic Venera design, including landers but also atmospheric balloons which relayed data for about two days.

## 1.22.2 Scientific findings

There were many scientific findings about Venus from the data retrieved by the Venera probes. For example, after analyzing the radar images returned from Venera 15 and 16, it was concluded that the ridges and grooves on the surface of Venus were the result of tectonic deformations.*[4]

## 1.22.3 Venera camera failures and success

The Venera 9 and 10 landers had two cameras each. Only one functioned because the lens covers failed to separate from the second camera on each lander. The design was changed for Venera 11 and 12, but this change made the problem worse and all cameras failed on those missions. Venera 13 and 14 were the only landers on which all cameras worked properly; although unfortunately, the titanium lens cap on Venera 14 landed precisely on the area which was targeted by the soil compression probe.

The external link at the bottom of the page shows all lander imagery.

## 1.22.4 Flight data for all Venera missions

## 1.22.5 See also

- Astron (spacecraft) – a space observatory derived from the Venera bus

- Pioneer Venus project

- Vega 1

- Vega 2

- Vega program

- Venera-D – First post-Soviet Venus probe (to be launched in 2024)

## 1.22.6 References

[1] Wade, Mark. "Venera 1VA". Encyclopedia Astronautica. Retrieved 28 July 2010.

[2] NSSDC Chronology of Venus Exploration (NASA Goddard Space Center), see also NSSDC Tentatively Identified (Soviet) Missions and Launch Failures (NASA Goddard Space Center), accessed August 9, 2010

[3] Ultimax Group's Venus Exploration Atlas page (accessed Aug 18 2010)

[4] Basilevsky, A. T.; Pronin, A. A.; Ronca, L. B.; Kryuchkov, V. P.; Sukhanov, A. L.; Markov, M. S. (1986). "Styles of tectonic deformations of Venus - Analysis of Venera 15 and 16 data (abstract only)". *Journal of Geophysical Research* (Journal of Geophysical Research March 30, 1986, p. D399-D411) **91**: 399. Bibcode:1986JGR....91..399B. doi:10.1029/JB091iB04p0D399. ISSN 0148-0227.

## 1.22.7 External links

- Media related to Venera program at Wikimedia Commons

- The Soviet Exploration of Venus

- Catalog of Soviet Venus images

- Missions to Venus by NASA's Solar System Exploration

- Venera 9, 10, 13, and 14 images of the surface of Venus

- The Soviets and Venus by Larry Klaes, 1993

# 1.23    Venera 1

**Venera 1** (Russian: Венера—1 meaning *Venus 1*), also known as **Venera-1VA No.2** and occasionally in the West as **Sputnik 8** was the first spacecraft to fly past Venus, as part of the Soviet Union's Venera programme.[*][1] Launched in February 1961, it flew past Venus on 19 May of the same year; however, radio contact with the probe was lost before the flyby, resulting in it returning no data.

## 1.23.1    Spacecraft

Venera 1 was a 643.5-kilogram (1,419 lb) probe consisting of a cylindrical body 1.05 metres (3 ft 5 in) in diameter topped by a dome, totalling 2.035 metres (6 ft 8.1 in) in height. This was pressurized to 1.2 standard atmospheres (120 kPa) with dry nitrogen, with internal fans to maintain even distribution of heat. Two solar panels extended from the cylinder, charging a bank of silver-zinc batteries. A 2-metre parabolic wire-mesh antenna was designed to send data from Venus to Earth on a frequency of 922.8 MHz. A 2.4-metre antenna boom was used to transmit short-wave signals during the near-Earth phase of the mission. Semidirectional quadrupole antennas mounted on the solar panels provided routine telemetry and telecommand contact with Earth during the mission, on a circularly-polarized decimetre radio band.

The probe was equipped with scientific instruments including a flux-gate magnetometer attached to the antenna boom, two ion traps to measure solar wind, micrometeorite detectors, and Geiger counter tubes and a sodium iodide scintillator for measurement of cosmic radiation. An experiment attached to one solar panel measured temperatures of experimental coatings. Infrared and/or ultraviolet radiometers may have been included. The dome contained a KDU-414 engine used for mid-course corrections. Temperature control was achieved by motorized thermal shutters.

During most of its flight, Venera 1 was spin stabilized. It was the first spacecraft designed to perform mid-course corrections, by entering a mode of 3-axis stabilization, fixing on the Sun and the star Canopus. Had it reached Venus, it would have entered another mode of 3-axis stabilization, fixing on the Sun and Earth, and using for the first time a parabolic antenna to relay data.

## 1.23.2    Launch

Venera 1 was the second of two attempts to launch a probe to Venus in February 1961, immediately following the launch of its sister ship Venera-1VA No.1,[*][2] which failed to leave Earth orbit.[*][3] Soviet experts launched Venera-1 using a Molniya carrier rocket from the Baikonur Cosmodrome. The launch took place at On 00:34:36 UTC on 12 February 1961.[*][4]

The spacecraft, along with the rocket's Blok-L upper stage, were initially placed into a 229 × 282 km low Earth orbit,[*][1] before the upper stage fired to place Venera 1 into a heliocentric orbit, directed towards Venus. The 11D33 engine was the world's first staged-combustion-cycle rocket engine, and also the first use of an ullage engine to allow a liquid-fuel rocket engine to start in space.

## 1.23.3    Failure

Three successful telemetry sessions were conducted, gathering solar-wind and cosmic-ray data near Earth, at the Earth's magnetopause, and on February 19 at a distance of 1,900,000 km. After discovering the solar wind with Luna 2, Venera 1 provided the first verification that this plasma was uniformly present in deep space. Seven days later, the next scheduled telemetry session failed to occur. On May 19, 1961, Venera 1 passed within 100,000 km of Venus. With the help of the British radio telescope at Jodrell Bank, some weak signals from Venera 1 may have been detected in June. Soviet engineers believed that Venera-1 failed due to the overheating of a solar-direction sensor.

## 1.23.4    See also

- Venera
- Mariner 2
- Timeline of planetary exploration

## 1.23.5    References

[1] NSSDC Spacecraft 1961-003A (NASA Goddard Space Center), accessed August 9, 2010

[2] NSSDC Chronology of Venus Exploration (NASA Goddard Space Center), accessed August 9, 2010

[3] NSSDC Tentatively Identified (Soviet) Missions and Launch Failures (NASA Goddard Space Center), accessed August 9, 2010

[4] McDowell, Jonathan.  "Launch Log" . *Jonathan's Space Page*. Retrieved 3 January 2013.

## 1.23.6    External links

- The Soviet Exploration of Venus
- http://nssdc.gsfc.nasa.gov/database/MasterCatalog? sc=1961-003A

# 1.24 Venera 7

The **Venera 7** (Russian: **Венера–7** meaning *Venus 7*) (manufacturer's designation: **3V (V-70)**) was a Soviet spacecraft, part of the Venera series of probes to Venus. When it landed on the Venusian surface, it became the first spacecraft to land on another planet and first to transmit data from there back to Earth.*[1]

## 1.24.1 Launch

The probe was launched from Earth on August 17, 1970, at 05:38 UTC. It consisted of an interplanetary bus based on the 3MV system and a lander.*[2] During the flight to Venus two in-course corrections were made using the bus's on-board KDU-414 engine.*[2]

## 1.24.2 Landing

It entered the atmosphere of Venus on December 15, 1970.*[2] The lander remained attached to the interplanetary bus during the initial stages of atmospheric entry*[2] to allow the bus to cool the lander to −8°C for as long as possible.*[2] The lander was ejected once atmospheric buffeting broke the interplanetary bus's lock-on with Earth.*[2] The parachute opened at a height of 60km and atmospheric testing began with results showing the atmosphere to be 97% carbon dioxide.*[2] The parachute appeared to fail during the descent, resulting in a descent more rapid than planned.*[2] As a result the lander struck the surface of Venus at about 16.5 metres per second (54 ft/s) at 05:37:10 UTC.*[2] Landing coordinates are 5°S 351°E / 5°S 351°E.*[3]

The probe appeared to go silent on impact*[2] but recording tapes kept rolling.*[4] A few weeks later, upon a review of the tapes, another 23 minutes of very weak signals were found on them.*[4] The spacecraft had landed on Venus and probably bounced onto its side, leaving the medium gain antenna not aimed correctly for strong signal transmission to Earth.*[4] The only data returned from the surface were temperature readings, which gave a temperature of 475 °C (887 °F).*[2]

## 1.24.3 See also

- Timeline of artificial satellites and space probes

## 1.24.4 References

[1] "Science: Onward from Venus". TIME. 8 February 1971. Retrieved 2 January 2013.

[2] Reeves, Robert (1994). *The Superpower Space Race: An Explosive Rivalry through the Solar System*. Plenum Press. pp. 211–215. ISBN 0-306-44768-1.

[3] Patrick Moore, *The data book of astronomy*. CRC Press, 2000, p. 92.
See Table 5-5, Missions to Venus, 1961-2000. Landing near Navka Planitia

[4] Larry Klaes, *THE SOVIETS AND VENUS, PART 1*, 1993.

## 1.24.5 External links

- Venera 7 NASA NSSDC Master Catalog Data
- Plumbing the Atmosphere of Venus

# 1.25 Venera 8

**Venera 8** (Russian: **Венера–8** meaning *Venus 8*) (manufacturer's designation: **3V (V-72)**) was a probe in the Soviet Venera program for the exploration of Venus.

Venera 8 was a Venus atmospheric probe and lander. Its instrumentation included temperature, pressure, and light sensors as well as an altimeter, gamma ray spectrometer, gas analyzer, and radio transmitters. The spacecraft took 117 days to reach Venus with one mid-course correction on 6 April 1972, separating from the bus (which contained a cosmic ray detector, solar wind detector, and ultraviolet spectrometer) and entering the atmosphere on 22 July 1972 at 08:37 UT. A refrigeration system attached to the bus was used to pre-chill the descent capsule's interior prior to atmospheric entry in order to prolong its life on the surface. Descent speed was reduced from 41,696 km/h to about 900 km/h by aerobraking. The 2.5 meter diameter parachute opened at an altitude of 60 km.

Venera 8 transmitted data during the descent. A sharp decrease in illumination was noted at 35 to 30 km altitude and wind speeds of less than 1 km/s were measured below 10 km. Venera 8 landed at 09:32 UT in what is now called Vasilisa Region, within 150 km radius of 10°42′S 335°15′E / 10.70°S 335.25°E, in sunlight, about 500 km from the morning terminator. The lander mass was 495 kg. It continued to send back data for 50 minutes, 11 seconds after landing before failing due to the harsh surface conditions. The probe confirmed the earlier data on the high Venus surface temperature and pressure (470 degrees Celsius, 90 atmospheres) returned by Venera 7, and also measured the light level as being suitable for surface photography, finding it to be similar to the amount of light on Earth on an overcast day with roughly 1 km visibility.

Venera 8's photometer measurements showed for the first time that the Venusian clouds end at a high altitude, and

*Venera 8 landing capsule*

the atmosphere was relatively clear from there down to the surface. The on-board gamma ray spectrometer measured the uranium/thorium/potassium ratio of the surface rock, indicating it was similar to granite.

### 1.25.1   Payload experiments

- Temperature and pressure sensors - ITD

- Accelerometer - DOU-1M

- Photometers - IOV-72

- Ammonia analyser - IAV-72

- Gamma ray spectrometer - GS-4

- Radar altimeter

- Radio Doppler experiment

### 1.25.2   See also

- Timeline of artificial satellites and space probes

### 1.25.3   External links

- Plumbing the Atmosphere of Venus

- Venera 8 NASA NSSDC Master Catalog Data

## 1.26   Venera 9

**Venera 9** (Russian: **Венера**−**9** meaning *Venus 9*), manufacturer's designation: **4V-1 No. 660**,[*][2] was a USSR unmanned space mission to Venus. It consisted of an orbiter and a lander. It was launched on June 8, 1975 02:38:00 UTC and had a mass of 4,936 kg (10,884 lb).[*][3] The orbiter was the first spacecraft to orbit Venus, while the lander was the first to return images from the surface of another planet.[*][4]

### 1.26.1   Orbiter

The orbiter consisted of a cylinder with two solar panel wings and a high gain parabolic antenna attached to the curved surface. A bell-shaped unit holding propulsion systems was attached to the bottom of the cylinder, and mounted on top was a 2.4 meter sphere which held the lander.

The orbiter entered Venus orbit on October 20, 1975. Its mission was to act as a communications relay for the lander and to explore cloud layers and atmospheric parameters with several instruments and experiments. It performed 17 survey missions from October 26, 1975 to December 25, 1975.

**List of orbiter instruments and experiments**

*180-degree panorama taken by Venera 9 of the surface of Venus*

[*][5]

- 1.6-2.8 μm IR Spectrometer

- 8-28 μm IR Radiometer

- 352 nm UV Photometer

- 2 Photo-polarimeters (335-800 nm)

- 300-800 nm Spectrometer

- . Lyman-α H/D Spectrometer

- Bistatic Radar Mapping

- CM, DM Radio Occultations

- Triaxial Magnetometer

- 345-380 nm UV Camera

- 355-445 nm Camera

- 6 Electrostatic Analyzers

- 2 Modulation Ion Traps

- Low-Energy Proton / Alpha detector

- Low-Energy Electron detector

- 3 Semiconductor Counters

- 2 Gas-Discharge Counters

- Cherenkov Detector

## 1.26.2   Lander

*Venera 9 lander*

On October 20, 1975, the lander spacecraft was separated from the orbiter, and landing was made with the Sun near zenith at 05:13 UTC on October 22.  Venera 9 landed within a 150 km radius of 31°01′N 291°38′E / 31.01°N 291.64°E, near Beta Regio, on a steep (20°) slope covered with boulders (suspected to be the slope of the tectonic rift valley, Aikhulu Chasma). The entry sphere weighed 1,560 kg (3,440 lb) and the surface payload 660 kg (1,455 lb).[*][6]

It was the first spacecraft to return an image from the surface of another planet.

A system of circulating fluid was used to distribute the heat load.  This system, plus pre-cooling prior to entry, permitted operation of the lander for 53 minutes after landing, at which time radio contact with the orbiter was lost as the orbiter moved out of radio range.[*][5]  During descent, heat dissipation and deceleration were accomplished sequentially by protective hemispheric shells, three parachutes, a disc-shaped drag brake, and a compressible, metal, doughnut-shaped landing cushion. The landing was about 2,200 km from the Venera 10 landing site.

Venera 9 measured clouds that were 30–40 km thick with bases at 30–35 km altitude. It also measured atmospheric chemicals including hydrochloric acid, hydrofluoric acid, bromine, and iodine.  Other measurements included surface pressure of about 90 atmospheres (9 MPa), temperature of 485 °C (905 °F), and surface light levels comparable to those at Earth mid-latitudes on a cloudy summer day.  Venera 9 was the first probe to send back black and white television pictures from the Venusian surface showing shadows, no apparent dust in the air, and a variety of 30 to 40 cm rocks which were not eroded. Planned 360-degree panoramic pictures could not be taken because one of two camera lens covers failed to come off, limiting pictures to 180 degrees. This failure recurred with Venera 10.

### Lander payload[*][5]

- Temperature and pressure sensors

- Accelerometer

- Visible / IR photometer - IOV-75

- Backscatter and multi-angle nephelometers - MNV-75

- P-11 Mass spectrometer - MAV-75

- Panoramic telephotometers (2, with lamps)

- Anemometer - ISV-75

- Gamma ray spectrometer - GS-12V

- Gamma ray densitometer - RP-75

## 1.26.3   Image processing

Don P. Mitchell recently came across the original Venera imaging data while researching the Soviet Venus program, and reconstructed the images using modern image processing software.

## 1.26.4   See also

- Timeline of artificial satellites and space probes

### 1.26.5   References

[1]  McDowell, Jonathan.  "Launch Log".  *Jonathan's Space Page*. Retrieved 11 April 2013.

[2]  RussianSpaceWeb.com

[3]  "NSSDC Master Catalog - Venera 9".  NASA National Science Data Center. Retrieved 13 April 2013.

[4]  Solar System Exploration Multimedia Gallery: Venera 9, NASA website, accessed August 7, 2009.

[5]  Mitchell, Don P.  "First Pictures of the Surface of Venus". Retrieved 13 April 2013.

[6]  Interplanetary Spacecraft

## 1.27   Venera 10

**Venera 10** (Russian: **Венера−10** meaning *Venus 10*), manufacturer's designation: **4V-1 No. 661**,[1] was a USSR unmanned space mission to Venus. It consisted of an orbiter and a lander. It was launched on June 14, 1975 03:00:31 UTC and had a mass of 5033 kg (11096 lb).[2]

### 1.27.1   Orbiter

The orbiter entered Venus orbit on October 23, 1975. Its mission was to serve as a communications relay for the lander and to explore cloud layers and atmospheric parameters with several instruments and experiments:[3]

- 1.6-2.8 μm IR Spectrometer

- 8-28 μm IR Radiometer

- 352 nm UV Photometer

- 2 Photopolarimeters (335-800 nm)

- 300-800 nm Spectrometer

- Lyman-α H/D Spectrometer

- Bistatic Radar Mapping

- CM, DM Radio Occultations

- Triaxial Magnetometer

- 345-380 nm UV Camera

- 355-445 nm Camera

- 6 Electrostatic Analyzers

- 2 Modulation Ion Traps

- Low-Energy Proton / Alpha detector

- Low-Energy Electron detector

- 3 Semiconductor Counters

- 2 Gas-Discharge Counters

- Cherenkov Detector

The orbiter consisted of a cylinder with two solar panel wings and a high gain parabolic antenna attached to the curved surface.   A bell-shaped unit holding propulsion systems was attached to the bottom of the cylinder, and mounted on top was a 2.4 meter sphere which held the landers.

### 1.27.2   Lander

*Venera 10 lander*

On October 23, 1975, this spacecraft was separated from the Orbiter, and landing was made with the sun near zenith, at 0517 UT, on October 25.

A system of circulating fluid was used to distribute the heat load. This system, plus precooling prior to entry, permitted operation of the spacecraft for 65 min after landing. During descent, heat dissipation and deceleration were accomplished sequentially by protective hemispheric shells, three

*Landing area of Venera 10 as mapped by the Magellan orbiter*

parachutes, a disk-shaped drag brake, and a compressible, metal, doughnut-shaped, landing cushion.[*][3]

It landed 2200 km from Venera 9 (within a 150 km radius of 15°25′N 291°31′E / 15.42°N 291.51°E), three days after its touchdown.[*][4] Venera 10 measured a surface wind-speed of 3.5 m/s. Other measurements included atmospheric pressure at various heights, and temperature, and surface light levels. Venera 10 was the second probe to send back black and white television pictures from the Venusian surface (after Venera 9). Venera 10 photographs showed lava rocks of pancake shape with lava or other weathered rocks in between. Planned 360 degree panoramic pictures could not be taken because, as with Venera 9, one of two camera lens covers failed to come off, limiting pictures to 180 degrees.

Lander Payload:[*][3]

- Temperature and pressure sensors
- Accelerometer
- Visible / IR photometer - IOV-75
- Backscatter and multi-angle nephelometers - MNV-75
- P-11 Mass spectrometer - MAV-75
- Panoramic telephotometers (2, with lamps)
- Anemometer - ISV-75
- Gamma ray spectrometer - GS-12V
- Gamma ray densitometer - RP-75
- Radio Doppler experiment

### 1.27.3 See also

- Timeline of artificial satellites and space probes

### 1.27.4 References

[1] Zak, Anatoly. "Venera-9 and 10". Russianspaceweb.com. Retrieved 14 April 2013.

[2] "NSSDC Master Catalog - Venera 10". NASA National Science Data Center. Retrieved 13 April 2013.

[3] Mitchell, Don P. "First Pictures of the Surface of Venus". Retrieved 13 April 2013.

[4] Interplanetary Spacecraft

## 1.28 Venera 11

The **Venera 11** (Russian: **Венера−11** meaning *Venus 11*) was a USSR unmanned space mission part of the Venera program to explore the planet Venus. Venera 11 was launched on 9 September 1978 at 3:25:39 UTC.[*][1]

Separating from its flight platform on December 23, 1978 the lander entered the Venus atmosphere two days later on December 25 at 11.2 km/s. During the descent, it employed aerodynamic braking followed by parachute braking and ending with atmospheric braking. It made a soft landing on the surface at 06:24 Moscow time (0324 UT) on 25 December after a descent time of approximately 1 hour. The touchdown speed was 7 to 8 m/s. Information was transmitted to the flight platform for retransmittal to earth until it moved out of range 95 minutes after touchdown.[*][2] Landing coordinates are 14°S 299°E / 14°S 299°E.[*][3]

### 1.28.1 Flight platform

After ejection of the lander probe, the flight platform continued on past Venus in a heliocentric orbit. Near encounter with Venus occurred on December 25, 1978, at approximately 34,000 km altitude. The flight platform acted as a data relay for the descent craft for 95 minutes until it flew out of range and returned its own measurements on interplanetary space.[*][4]

Venera 11 flight platform carried solar wind detectors, ionosphere electron instruments and two gamma ray burst detectors – the Soviet-built KONUS and the French-built SIGNE 2. The SIGNE 2 detectors were simultaneously flown on Venera 12 and Prognoz 7 to allow triangulation of gamma ray sources. Before and after Venus flyby, Venera 11 and Venera 12 yielded detailed time-profiles for 143 gamma-ray bursts, resulting in the first ever catalog of such events.

The last gamma-ray burst reported by Venera 11 occurred on January 27, 1980

List of flight platform instruments and experiments:[*][5]

- 30–166 nm Extreme UV Spectrometer
- Compound Plasma Spectrometer
- KONUS Gamma-Ray Burst Detector
- SNEG Gamma-Ray Burst Detector
- Magnetometer
- 4 Semiconductor Counters
- 2 Gas-Discharge Counters
- 4 Scintillation Counters
- Hemispherical Proton Telescope

The mission ended in February, 1980.

## 1.28.2 Lander

The lander carried instruments to study the temperature and atmospheric and soil chemical composition. A device called Groza detected lightning on Venus. Both Venera 11 and Venera 12 had landers with two cameras, each designed for color imaging, though Soviet literature does not mention them. Each failed to return images when the lens covers did not separate after landing due to a design flaw. The soil analyzer also failed. A gas chromatograph was on board to measure the composition of the Venus atmosphere, as well as instruments to study scattered solar radiation. Results reported included evidence of lightning and thunder, a high $Ar^{36}/Ar^{40}$ ratio, and the discovery of carbon monoxide at low altitudes.[*][2]

List of lander experiments and instruments:

- Backscatter Nephelometer
- Mass Spectrometer – MKh-6411
- Gas Chromatograph – Sigma
- X-Ray Fluorospectrometer
- 360° Scanning Photometer – IOAV
- Spectrometer (430–1170 nm)
- Microphone/Anemometer
- Low-Frequency Radio Sensor
- 4 Thermometers

- 3 Barometers
- Accelerometer – Bizon
- Penetrometer – PrOP-V
- Soil Analysis Device
- 2 Color Cameras
- Small solar batteries – MSB

## 1.28.3 See also

- Timeline of artificial satellites and space probes

## 1.28.4 References

[1] "Venera 11" .

[2] "Venera 11 Descent Craft" .

[3] "Venera 11 – Detail" .

[4] "Venera 11 (NASA NSS-DC)".

[5] Mitchell, Don P. "Drilling into the Surface of Venus" . Retrieved 13 April 2013.

## 1.28.5 External links

- Venera 11 & Venera 12 (NASA)
- Experiments on Venera 11 (NASA NSS-DC) Has detail on each experiment/instrument.
- Drilling into the Surface of Venus (Venera 11 and 12)

# 1.29 Venera 12

The **Venera 12** (Russian: **Венера—12** meaning *Venus 12*) was an USSR unmanned space mission to explore the planet Venus. Venera 12 was launched on 14 September 1978 at 02:25:13 UTC.[*][1]

Separating from its flight platform on December 19, 1978, the lander entered the Venus atmosphere two days later at 11.2 km/s. During the descent, it employed aerodynamic braking followed by parachute braking and ending with atmospheric braking. It made a soft landing on the surface at 06:30 Moscow time (0330 UT) on 21 December after a descent time of approximately 1 hour. The touchdown speed was 7–8 m/s. Landing coordinates are 7°S 294°E / 7°S 294°E. It transmitted data to the flight platform for 110 minutes after touchdown until the flight platform moved out of range while remaining in a heliocentric orbit. Identical instruments were carried on Venera 11 and 12.[*][2]

## 1.29.1 Flight platform

Venera 12 flight platform carried solar wind detectors, ionosphere electron instruments and two gamma ray burst detectors – the Soviet-built KONUS and the French-built SIGNE 2. The SIGNE 2 detectors were simultaneously flown on Venera 12 and Prognoz 7 to allow triangulation of gamma ray sources. Before and after Venus flyby, Venera 11 and Venera 12 yielded detailed time-profiles for 143 gamma-ray bursts, resulting in the first ever catalog of such events. The last gamma-ray burst reported by Venera 12 occurred on January 5, 1980. Venera 12 used its ultraviolet spectrometer to study Comet Bradfield on 13 February 1980, and reported spectrophotometric data until 19 March 1980.[*][3]

List of flight platform instruments and experiments:[*][4]

- 30–166 nm Extreme UV Spectrometer

- Compound Plasma Spectrometer

- KONUS Gamma-Ray Burst Detector

- SNEG Gamma-Ray Burst Detector

- Magnetometer

- 4 Semiconductor Counters

- 2 Gas-Discharge Counters

- 4 Scintillation Counters

- Hemispherical Proton Telescope

The active phase of the science mission for the flight platform ended in April, 1980.

## 1.29.2 Lander

The Venera 12 descent craft carried instruments designed to study the detailed chemical composition of the atmosphere, the nature of the clouds, and the thermal balance of the atmosphere. Among the instruments on board was a gas chromatograph to measure the composition of the Venus atmosphere, instruments to study scattered solar radiation and soil composition, and a device named Groza which was designed to measure atmospheric electrical discharges. Results reported included evidence of lightning and thunder, a high $Ar^{36}/Ar^{40}$ ratio, and the discovery of carbon monoxide at low altitudes. Both Venera 11 and Venera 12 had landers with two cameras, each designed for color imaging. Each failed to return images when the lens covers did not separate after landing due to a design flaw.[*][5]

List of lander experiments and instruments:[*][4]

- Backscatter Nephelometer

- Mass Spectrometer – MKh-6411

- Gas Chromatograph – Sigma

- X-Ray Fluorospectrometer

- 360° Scanning Photometer – IOAV

- Spectrometer (430–1170 nm)

- Microphone/Anemometer

- Low-Frequency Radio Sensor

- 4 Thermometers

- 3 Barometers

- Accelerometer – Bizon

- Penetrometer – PrOP-V

- Soil Analysis Device

- 2 Color Cameras

- Small solar batteries – MSB

## 1.29.3 See also

- Timeline of artificial satellites and space probes

## 1.29.4 References

[1] "Venera 12" .

[2] "Venera 12 – Detail" .

[3] "Venera 12 (NASA NSS-DC)".

[4] Mitchell, Don P. "Drilling into the Surface of Venus" . Retrieved 13 April 2013.

[5] "Venera 12 Descent Craft" .

## 1.29.5 External links

- Venera 11 & Venera 12 (NASA)

- Experiments on Venera 12 (NASA NSS-DC) Has detail on each experiment/instrument.

- Drilling into the Surface of Venus (Venera 11 and 12)

# 1.30    Venera 13

**Venera 13** (Russian: **Венера−13** meaning *Venus 13*) was a probe in the Soviet Venera program for the exploration of Venus.

Venera 13 and 14 were identical spacecraft built to take advantage of the 1981 Venus launch opportunity and launched 5 days apart, Venera 13 on 30 October 1981 at 06:04:00 UTC and Venera 14 on 4 November 1981 at 05:31:00 UTC, both with an on-orbit dry mass of 760 kg.

## 1.30.1    Design

Each mission consisted of a cruise stage and an attached descent craft.

### Cruise stage

As the cruise stage flew by Venus the bus acted as a data relay for the lander and then continued on into a heliocentric orbit. It was equipped with a gamma-ray spectrometer, UV grating monochromator, electron and proton spectrometers, gamma-ray burst detectors, solar wind plasma detectors, and two-frequency transmitters which made measurements before, during, and after the Venus flyby.

### Descent lander

The descent lander was an hermetically-sealed pressure vessel, which contained most of the instrumentation and electronics, mounted on a ring-shaped landing platform and topped by an antenna. The design was similar to the earlier Venera 9–12 landers. It carried instruments to take chemical and isotopic measurements, monitor the spectrum of scattered sunlight, and record electric discharges during its descent phase through the Venusian atmosphere. The spacecraft utilized a camera system, an X-ray fluorescence spectrometer, a screw drill and surface sampler, a dynamic penetrometer, and a seismometer to conduct investigations on the surface.

List of lander experiments and instruments:

- Accelerometer, Impact Analysis - Bison-M

- Thermometers, Barometers - ITD

- Spectrometer / Directional Photometer - IOAV-2

- Ultraviolet Photometer

- Mass Spectrometer - MKh-6411

- Penetrometer / Soil Ohmmeter - PrOP-V

- Chemical Redox Indicator - Kontrast

- 2 Color Telephotometer Cameras - TFZL-077

- Gas Chromatograph - Sigma-2

- Radio / Seismometer - Groza-2

- Nephelometer - MNV-78-2

- Hydrometer - VM-3R

- X-Ray Fluorescence Spectrometer (Aerosol) - BDRA-1V

- X-Ray Fluorescence Spectrometer (Soil) - Arakhis-2

- Soil Drilling Apparatus - GZU VB-02

- Stabilized Oscillator / Doppler Radio

- Small solar batteries - MSB

## 1.30.2    Landing

After launch and a four-month cruise to Venus the descent vehicle separated from the cruise stage and plunged into the Venusian atmosphere on 1 March 1982. After entering the atmosphere a parachute was deployed. At an altitude of about 50 km the parachute was released and simple air-braking was used the rest of the way to the surface.

Venera 13 landed at 7°30′S 303°00′E / 7.5°S 303°E, about 950 km northeast of Venera 14, just east of the eastern extension of an elevated region known as Phoebe Regio.[1]

The lander had cameras to take pictures of the ground and spring-loaded arms to measure the compressibility of the soil. The quartz camera windows were covered by lens caps which popped off after descent.[2][3]

The area was composed of bedrock outcrops surrounded by dark, fine-grained soil. After landing, an imaging panorama was started and a mechanical drilling arm reached to the surface and obtained a sample, which was deposited in a hermetically sealed chamber, maintained at 30 °C and a pressure of about 0.05 atmosphere (5 kPa). The composition of the sample determined by the X-ray fluorescence spectrometer put it in the class of weakly differentiated melanocratic alkaline gabbroids.

The lander functioned for 127 minutes (the planned design life was 32 minutes) in an environment with a temperature of 457 °C (855 °F) and a pressure of 89 Earth atmospheres (9.0 MPa). The descent vehicle transmitted data to the satellite, which acted as a data relay as it flew by Venus.

### 1.30.3 Suggested photographic evidence of life

Leonid Ksanfomaliti of the Space Research Institute of Russia's Academy of Sciences (a contributor to the Venera mission) and Stan Karaszewski of Karas, in *Solar System Research*, suggests signs of life in the Venera images.*[4] According to Ksanfomaliti, certain objects resembled a "disk", a "black flap" and a "scorpion" which "emerge, fluctuate and disappear", referring to their changing location on different photographs and traces on the ground.*[5]*[6]*[7] The article was titled "Venus as a natural laboratory for search of life in high temperature conditions: Events on the planet on March 1, 1982", L. V. Ksanfomality, Solar System Research, February 2012, Volume 46, Issue 1, pp 41–5. Engineers familiar with the probe have identified the moving "disk" as actually being the two lens caps ejected from the lander. Rather than a single object that had moved between two different places, they are simply two inanimate similar-looking objects in different places. The other "objects" are ascribed to image processing artifacts and do not appear in the original photography.*[8]

The editors of that journal published an editorial comment and a number of commentary articles from other scientists in the September 2012 issue, Volume 46, Issue 5. That issue also includes a second article from Ksanfomaliti, in which he identifies several other life forms and speculates regarding the apparent rich diversity of life around the landing site.

### 1.30.4 In fiction

- The *Venera 13* lander appears in the short film *Horses on Mars* (2001) with a message to the main character who is a microbe lost on Venus.

### 1.30.5 References

[1] "NSSDC Master Catalog - Venera 13 Descent Craft". NASA National Science Data Center. Retrieved 13 April 2013.

[2] Dr Karl - Murphy's Law, Part two

[3] Images available at http://www.donaldedavis.com/2003NEW/NEWSTUFF/DDVENUS.html

[4] Solar System Research, 2012, 46(1)

[5] "Russian Researcher Suggests Venera-13 Imaged Life on Venus | Space Exploration". Sci-News.com. 1982-03-01. Retrieved 2012-08-12.

[6] Space. "Russian scientist claims 1982 pictures shows 'life on Venus'". Telegraph. Retrieved 2012-08-12.

[7] "Scientists Move on from Mars, Say Venus Shows Signs of Life - International Business Times". Ibtimes.com. 2012-01-23. Retrieved 2012-08-12.

[8] http://www.pcmag.com/article2/0,2817,2399294,00.asp

### 1.30.6 External links

- Soviet Venus Images

- Venera: The Soviet Exploration of Venus

- On Ksanfomality's Venus Life Result

## 1.31 Venera 14

**Venera 14** (Russian: **Венера–14** meaning *Venus 14*) was a probe in the Soviet Venera program for the exploration of Venus.

Venera 14 was identical to the Venera 13 spacecraft and built to take advantage of the 1981 Venus launch opportunity and launched 5 days apart. It was launched on 4 November 1981 at 05:31:00 UTC and Venera 13 on 30 October 1981 at 06:04:00 UTC, both with an on-orbit dry mass of 760 kg.

### 1.31.1 Design

Each mission consisted of a cruise stage and an attached descent craft.

**Cruise stage**

As the cruise stage flew by Venus the bus acted as a data relay for the lander and then continued on into a heliocentric orbit. It was equipped with a gamma-ray spectrometer, UV grating monochromator, electron and proton spectrometers, gamma-ray burst detectors, solar wind plasma detectors, and two-frequency transmitters which made measurements before, during, and after the Venus flyby.

**Descent lander**

The descent lander was a hermetically sealed pressure vessel, which contained most of the instrumentation and electronics, mounted on a ring-shaped landing platform and topped by an antenna. The design was similar to the earlier Venera 9–12 landers. It carried instruments to take chemical and isotopic measurements, monitor the spectrum of scattered sunlight, and record electric discharges during its descent phase through the Venusian atmosphere. The spacecraft utilized a camera system, an X-ray fluorescence spectrometer, a screw drill and surface sampler, a dynamic penetrometer, and a seismometer to conduct investigations on the surface.

List of lander experiments and instruments:*[1]

- Accelerometer, Impact Analysis - Bison-M

- Thermometers, Barometers - ITD

- Spectrometer / Directional Photometer - IOAV-2

- Ultraviolet Photometer

- Mass Spectrometer - MKh-6411

- Penetrometer / Soil Ohmmeter - PrOP-V

- Chemical Redox Indicator - Kontrast

- 2 Color Telephotometer Cameras - TFZL-077

- Gas Chromatograph - Sigma-2

- Radio / Microphone / Seismometer - Groza-2

- Nephelometer - MNV-78-2

- Hydrometer - VM-3R

- X-Ray Fluorescence Spectrometer (Aerosol) - BDRA-1V

- X-Ray Fluorescence Spectrometer (Soil) - Arakhis-2

- Soil Drilling Apparatus - GZU VB-02

- Stabilized Oscillator / Doppler Radio

- Small solar batteries - MSB

### 1.31.2   Landing

After launch and a four-month cruise to Venus the descent vehicle separated from the bus and plunged into the Venusian atmosphere on March 5, 1982. After entering the atmosphere a parachute was deployed. At an altitude of about 50 km the parachute was released and simple airbraking was used the rest of the way to the surface.

Venera 14 landed at 13°15′S 310°00′E / 13.25°S 310°E (about 950 km southwest of Venera 13) near the eastern flank of Phoebe Regio on a basaltic plain.

The lander had cameras to take pictures of the ground and spring-loaded arms to measure the compressibility of the soil. The quartz camera windows were covered by lens caps which popped off after descent. Venera 14, however, ended up measuring the compressibility of the lens cap, which landed right where the probe was to measure the soil.*[2]

The composition of the surface samples was determined by the X-ray fluorescence spectrometer, showing it to be similar to oceanic tholeiitic basalts.

The lander survived 57 minutes (the planned design life was 32 minutes) in an environment with a temperature of 465 °C (869 °F) and a pressure of 94 Earth atmospheres (9.5 MPa). Telemetry had been maintained by means of the orbiting bus that carried signals from the lander's uplink antenna.*[3]

### 1.31.3   Fictional references

- *Venera 14* is visited by a Russian cosmonaut in BBC's *Space Odyssey: Voyage To The Planets*.

### 1.31.4   Image processing

American researcher Don P. Mitchell has processed the color images from Venera 13 and 14 using the raw original data.*[4] The new images are based on a more accurate linearization of the original 9-bit logarithmic pixel encoding.

### 1.31.5   References

[1] Mitchell, Don P. "Drilling into the Surface of Venus". Retrieved 13 April 2013.

[2] Images available at http://www.donaldedavis.com/2003NEW/NEWSTUFF/DDVENUS.html

[3] http://nssdc.gsfc.nasa.gov/nmc/spacecraftDisplay.do?id=1981-110D

[4] The versions currently available on Mitchell's website

### 1.31.6   External links

## 1.32   Venera 15

**Venera 15** (Russian: Венера−15 meaning *Venus 15*) was a spacecraft sent to Venus by the Soviet Union. This unmanned orbiter was to map the surface of Venus using high resolution imaging systems. The spacecraft was identical to Venera 16 and based on modifications to the earlier Venera space probes.

### 1.32.1   Mission profile

Venera 15 was launched on June 2, 1983 at 02:38:39 UTC and reached Venus' orbit on October 10, 1983.

The spacecraft was inserted into Venus orbit a day apart from Venera 16, with its orbital plane shifted by an angle of approximately 4° relative to one another probe. This made it possible to reimage an area if necessary. The spacecraft was in a nearly polar orbit with a periapsis ~1000 km, at 62°N latitude, and apoapsis ~65000 km, with an inclination ~90°, the orbital period being ~24 hours.

Together with Venera 16, the spacecraft imaged the area from the north pole down to about 30°N latitude (i.e. approx. 25% of Venus surface) over the 8 months of mapping operations.

### 1.32.2 Spacecraft structure

The Venera 15 and 16 spacecraft were identical and were based on modifications to the orbiter portions of the Venera 9 and Venera 14 probes. Each spacecraft consisted of a 5 m long cylinder with a 0.6 m diameter, 1.4 m tall parabolic dish antenna for the synthetic aperture radar (SAR) at one end. A 1-meter diameter parabolic dish antenna for the radio altimeter was also located at this end. The electrical axis of the radio altimeter antenna was lined up with the axis of the cylinder. The electrical axis of the SAR deviated from the spacecraft axis by 10 degrees. During imaging, the radio altimeter would be lined up with the center of the planet (local vertical) and the SAR would be looking off to the side at 10 degrees. A bulge at the opposite end of the cylinder held fuel tanks and propulsion units. Two square solar arrays extended like wings from the sides of the cylinder. A 2.6 m radio dish antenna for communications was also attached to the side of the cylinder. The spacecraft each massed 4000 kg.

Both Venera 15 and 16 were equipped with a Synthetic Aperture Radar (SAR). A radar was necessary in this mission because nothing else would be able to penetrate the dense clouds of Venus. The probes were equipped with on board computers that saved the images until the entire image was complete. This radar system replaced the normal landers that previous Venera probes brought to Venus.

List of spacecraft instruments and experiments:

- Polyus-V Synthetic Aperture Radar
- Omega Radar Altimeter
- Infrared Fourier Spectrometer
- Cosmic-Ray Detectors (6 sensors)
- Solar-Plasma Detectors

### 1.32.3 External links

- The Soviet Exploration of Venus

- Catalog of Soviet Venus images
- Venera 15
- Venera 15/16 Radar Mosaic Browser

## 1.33 Venera 16

**Venera 16** (Russian: Венера—16 meaning *Venus 16*) was a spacecraft sent to Venus by the Soviet Union. This unmanned orbiter was to map the surface of Venus using high resolution imaging systems. The spacecraft was identical to Venera 15 and based on modifications to the earlier Venera space probes.

### 1.33.1 Mission profile

Venera 16 was launched on June 7, 1983 at 02:32:00 UTC and reached Venus' orbit on October 11, 1983.

The spacecraft was inserted into Venus orbit a day apart from Venera 15, with its orbital plane shifted by an angle of approximately 4° relative to one another probe. This made it possible to reimage an area if necessary. The spacecraft was in a nearly polar orbit with a periapsis ~1000 km, at 62°N latitude, and apoapsis ~65000 km, with an inclination ~90°, the orbital period being ~24 hours.

Together with Venera 15, the spacecraft imaged the area from the north pole down to about 30°N latitude (i.e. approx. 25% of Venus surface) over the 8 months of mapping operations.

### 1.33.2 Spacecraft structure

The Venera 15 and 16 spacecraft were identical and were based on modifications to the orbiter portions of the Venera 9 and Venera 14 probes. Each spacecraft consisted of a 5 m long cylinder with a 0.6 m diameter, 1.4 m tall parabolic dish antenna for the synthetic aperture radar (SAR) at one end. A 1-meter diameter parabolic dish antenna for the radio altimeter was also located at this end. The electrical axis of the radio altimeter antenna was lined up with the axis of the cylinder. The electrical axis of the SAR deviated from the spacecraft axis by 10 degrees. During imaging, the radio altimeter would be lined up with the center of the planet (local vertical) and the SAR would be looking off to the side at 10 degrees. A bulge at the opposite end of the cylinder held fuel tanks and propulsion units. Two square solar arrays extended like wings from the sides of the cylinder. A 2.6 m radio dish antenna for communications was also attached to the side of the cylinder. The spacecraft each massed 4000 kg.

Both Venera 15 and 16 were equipped with a Synthetic Aperture Radar (SAR). A radar was necessary in this mission because nothing else would be able to penetrate the dense clouds of Venus. The probes were equipped with on board computers that saved the images until the entire image was complete.

List of spacecraft instruments and experiments:

- Polyus-V Synthetic Aperture Radar

- Omega Radar Altimeter

- Infrared Fourier Spectrometer

- Cosmic-Ray Detectors (6 sensors)

- Solar-Plasma Detectors

### 1.33.3   External links

- The Soviet Exploration of Venus

- Catalog of Soviet Venus images

- Venera 16

- Venera 15/16 Radar Mosaic Browser

## 1.34   Venera 2MV-1 No.2

**Venera 2MV-1 No.2**,[1][2] also known as **Sputnik 20** in the West, was a Soviet spacecraft, which was launched in 1962 as part of the Venera programme, and was intended to become the first spacecraft to land on Venus.[3] Due to a problem with its upper stage it failed to leave low Earth orbit, and reentered the atmosphere a few days later.[4] It was the second of two Venera 2MV-1 spacecraft, both of which failed to leave Earth orbit. The previous mission, Venera 2MV-1 No.1, was launched several days earlier.[2]

Venera 2MV-1 No.2 was launched at 02:12:30 UTC on 1 September 1962, atop a Molniya 8K78 carrier rocket flying from Site 1/5 at the Baikonur Cosmodrome.[1] The lower stages of the rocket operated nominally, injecting the fourth stage and payload into a low Earth orbit. Following a coast phase, the upper stage was to have ignited around sixty-one minutes and thirty seconds after launch, in order to place the spacecraft into heliocentric orbit. The ignition command did not reach the engine however, and the fuel valves did not open, so the upper stage failed to ignite leaving the payload in geocentric orbit.[5] It reentered the atmosphere on 6 September 1962, five days after it had been launched.[6]

The designations Sputnik 24, and later Sputnik 20 were used by the United States Naval Space Command to identify the spacecraft in its Satellite Situation Summary documents, since the Soviet Union did not release the internal designations of its spacecraft at that time, and had not assigned it an official name due to its failure to depart Earth orbit.[3][7] [8]

### 1.34.1   References

[1] McDowell, Jonathan. "Launch Log". Jonathan's Space Page. Retrieved 28 July 2010.

[2] Krebs, Gunter. "Venera (2a), (2b) (2MV-1 #1, 2)". Gunter's Space Page. Retrieved 28 July 2010.

[3] Zak, Anatoly. "Russia's unmanned missions to Venus". RussianSpaecWeb. Retrieved 28 July 2010.

[4] Wade, Mark. "Venera". Encyclopedia Astronautica. Retrieved 28 July 2010.

[5] Wade, Mark. "Soyuz". Encyclopedia Astronautica. Retrieved 28 July 2010.

[6] McDowell, Jonathan. "Satellite Catalog". Jonathan's Space Page. Retrieved 28 July 2010.

[7] Robbins, Stuart J. (11 January 2006). "Soviet Craft - Sputnik". Journey Through The Galaxy. Retrieved 28 July 2010.

[8] "Sputnik 20". NASA NSSDC. Retrieved 28 July 2010.

## 1.35   Venera 2MV-2 No.1

**Venera 2MV-2 No.1**,[1][2] also known as **Sputnik 21** in the West, was a Soviet spacecraft, which was launched in 1962 as part of the Venera programme, and was intended to make a flyby of Venus.[3] Due to a problem with the rocket which launched it, it failed to leave low Earth orbit, and reentered the atmosphere a few days later.[4] It was the second Venera 2MV-2 spacecraft, both of which failed to leave Earth orbit.[2]

Venera 2MV-2 No.1 was launched at 00:59:13 UTC on 12 September 1962, atop a Molniya 8K78 carrier rocket flying from Site 1/5 at the Baikonur Cosmodrome.[1] The rocket performed nominally up until cutoff of the Blok I stage, following injection into a low Earth orbit. Following cutoff, one of the oxidiser valves failed to close, and liquid oxygen was allowed to flow into the combustion chamber of one of the vernier thrusters. The vernier thruster exploded,[5] causing the rocket to tumble out of control. This led to the formation of bubbles in the upper stage oxidiser pump, which caused the upper stage engine to fail less than a second after ignition.[2] It reentered the atmosphere on 14 September 1962, two days after it had been launched.[6]

The designations Sputnik 25, and later Sputnik 21 were used by the United States Naval Space Command to identify the spacecraft in its Satellite Situation Summary documents, since the Soviet Union did not release the internal designations of its spacecraft at that time, and had not assigned it an official name due to its failure to depart geocentric orbit.[*][3][*][7][*][8]

### 1.35.1 References

[1] McDowell, Jonathan. "Launch Log". Jonathan's Space Page. Retrieved 28 July 2010.

[2] Krebs, Gunter. "Venera (2c) (2MV-2 #1)". Gunter's Space Page. Retrieved 28 July 2010.

[3] Zak, Anatoly. "Russia's unmanned missions to Venus". RussianSpaecWeb. Retrieved 28 July 2010.

[4] Wade, Mark. "Venera". Encyclopedia Astronautica. Retrieved 28 July 2010.

[5] Wade, Mark. "Soyuz". Encyclopedia Astronautica. Retrieved 28 July 2010.

[6] McDowell, Jonathan. "Satellite Catalog". Jonathan's Space Page. Retrieved 28 July 2010.

[7] Robbins, Stuart J. (11 January 2006). "Soviet Craft - Sputnik". Journey Through The Galaxy. Retrieved 28 July 2010.

[8] "Sputnik 21". NASA NSSDC. Retrieved 28 July 2010.

## 1.36 Venera-D

The **Venera-D** (Russian: Венера-Д, pronounced [vʲɪˈnʲɛrə ˈdɛ]) probe is a proposed Russian space probe to Venus, to be launched around 2025.[*][1][*][2] Venera-D's prime purpose is to make radar remote-sensing observations around the planet Venus in a manner similar to that of the Venera 15 and Venera 16 probes in the 1980s or the U.S. Magellan in the 1990s, but with the use of more-powerful radar. Venera-D is also intended to map future landing sites. A lander, based on the Venera design, is also planned, capable of surviving for a long duration on the planet's surface.

Venera-D will be the first Venus probe launched by the Russian Federation (the earlier Venera probes were launched by the former Soviet Union). Venera-D will serve as the flagship for a new generation of Russian-built Venus probes, culminating with a lander capable of withstanding the harsh Venerian environment for more than the 1½ hours logged by the Soviet probes. In order to keep research and development costs down, the new Venera-D probe will most likely resemble the Soviet probes, but will rely on new technologies developed by Russia since its last Venus missions

(Vega 1 and Vega 2 in 1985). Venera-D will most likely be launched on the Proton booster, but may be designed to be launched on the more powerful Angara rocket instead.

### 1.36.1 History

In 2003, Venera-D was proposed to the Russian Academy of Sciences for its "wish list" of scientific projects to be included into the Federal Space Program in 2006–2015. During the formulation of the mission concept in 2004, the launch of Venera-D was expected in 2013 and its landing on the surface of Venus in 2014.[*][3] In its original conception, it had a large orbiter, sub-satellite, two balloons, two small landers, and a large, long-lived lander. By 2011, the mission had been pushed back to 2018, and scaled back to an orbiter with a subsatellite orbiter, and a single lander with an expected 3 hour lifetime.[*][4] By the beginning of 2011, the Venera-D project entered Phase A (Preliminary Design) stage of development. Following the loss of the Phobos-Grunt spacecraft in November 2011 and resulting delays in all Russian planetary projects, the implementation of the project was again delayed to no earlier than 2025.[*][1]

### 1.36.2 References

[1] Foust, Jeff (16 November 2015). "Ben Carson's Prescription for Space Exploration". *Space News*. Retrieved 2015-12-07.

[2] "РАН: запуск "Венеры-Д" состоится не ранее 2024 года". Gazieta.ru. 9 April 2012. Retrieved September 26, 2013.

[3] Venera-D mission at Russia Space Web (accessed 25 November 2013)

[4] Ted Stryk, Russia's Venera-D mission (DPS-EPSC 2011), Planetary Society, 10/05/2011 (accessed 25 November 2013)

### 1.36.3 External links

- Venera-D – Federal Space program of Russian Federation

- Venera-D mission at Russia Space Web

## 1.37 Venus Express

*Venus Express* (**VEX**) was the first Venus exploration mission of the European Space Agency (ESA). Launched in November 2005, it arrived at Venus in April 2006 and began continuously sending back science data from its polar orbit around Venus. Equipped with seven scientific instruments, the main objective of the mission was the long term observation of the Venusian atmosphere. The observation over such long periods of time had never been done in previous missions to Venus, and was key to a better understanding of the atmospheric dynamics. It was hoped that such studies can contribute to an understanding of atmospheric dynamics in general, while also contributing to an understanding of climate change on Earth. ESA concluded the mission in December 2014.[*][4]

### 1.37.1    History

The mission was proposed in 2001 to reuse the design of the *Mars Express* mission. However, some mission characteristics led to design changes: primarily in the areas of thermal control, communications and electrical power. For example, since Mars is approximately twice as far from the Sun as Venus is, the radiant heating of the spacecraft is four times greater for *Venus Express* than *Mars Express*. Also, the ionizing radiation environment is harsher. On the other hand, the more intense illumination of the solar panels results in more generated photovoltaic power. The *Venus Express* mission also uses some spare instruments developed for the *Rosetta* spacecraft. The mission was proposed by a consortium led by D. Titov (Germany), E. Lellouch (France) and F. Taylor (United Kingdom).

The launch window for *Venus Express* was open from 26 October to 23 November 2005, with the launch initially set for 26 October 4:43 UTC. However, problems with the insulation from the Fregat upper stage led to a two-week launch delay to inspect and clear out the small insulation debris that migrated on the spacecraft.[*][5] It was eventually launched by a Soyuz-FG/Fregat rocket from the Baikonur Cosmodrome in Kazakhstan on 9 November 2005 at 03:33:34 UTC into a parking Earth orbit and 1 h 36 min after launch put into its transfer orbit to Venus. A first trajectory correction maneuver was successfully performed on 11 November 2005. It arrived at Venus on 11 April 2006, after 153 days of journey, and fired its main engine between 07:10:29 and 08:00:42 UTC SCET to reduce its velocity so that it could be captured by Venusian gravity into a nine-day orbit of 400 by 330,000 kilometres (250 by 205,050 mi).[*][6] The burn was monitored from ESA's Control Centre, ESOC, in Darmstadt, Germany.

Seven further orbit control maneuvers, two with the main engine and five with the thrusters, were required for *Venus Express* to reach its final operational 24-hour orbit around Venus.[*][6]

*Venus Express* entered its target orbit at apoapsis on 7 May 2006 at 13:31 UTC, when the spacecraft was 151,000,000 kilometres (94,000,000 mi) from Earth. At this point the

spacecraft was running on an ellipse substantially closer to the planet than during the initial orbit. The polar orbit ranged between 250 and 66,000 kilometres (160 and 41,010 mi) over Venus. The periapsis is located almost above the North pole (80° North latitude), and it takes 24 hours for the spacecraft to travel around the planet.

*Venus Express* studied the Venusian atmosphere and clouds in detail, the plasma environment and the surface characteristics of Venus from orbit. It is also made global maps of the Venusian surface temperatures. Its nominal mission was originally planned to last for 500 Earth days (approximately two Venusian sidereal days), but the mission has been extended three times: first on 28 February 2007 until early May 2009; then on 4 February 2009 until 31 December 2009; and then on 7 October 2009 until 31 December 2012.[*][7]

On 22 November 2010, the mission was extended to 2014.[*][8] On 20 June 2013, the mission was extended until 2015.[*][9] There were on-board resources for an additional 500 Earth days.

On 28 November 2014, mission control lost contact with *Venus Express*. Intermittent contact was reestablished on 3 December 2014, though there was no control over the spacecraft, likely due to exhaustion of propellant.[*][10] On 16 December 2014, ESA announced that the *Venus Express* mission had ended.[*][4] A carrier signal was still being received from the vehicle, but no data was being transmitted. Mission manager Patrick Martin expected the spacecraft would fall below 150 kilometres (93 mi) in early January 2015, with destruction occurring in late January or early February.[*][11] The spacecraft's carrier signal was last detected by ESA on 18 January 2015.[*][1]

### 1.37.2    Instruments

**ASPERA-4**: An acronym for "**A**nalyzer of **S**pace **P**lasmas and **E**nergetic **A**toms," ASPERA-4 investigated the interaction between the solar wind and the Venusian atmosphere, determine the impact of plasma processes on the atmosphere, determine global distribution of plasma and neutral gas, study energetic neutral atoms, ions and electrons, and analyze other aspects of the near Venus environment. ASPERA-4 is a re-use of the ASPERA-3 design used on *Mars Express*, but adapted for the harsher near-Venus environment.

**VMC**: The **V**enus **M**onitoring **C**amera is a wide-angle, multi-channel CCD. The VMC is designed for global imaging of the planet.[*][12] It operates in the visible, ultraviolet, and near infrared spectral ranges, and maps surface brightness distribution searching for volcanic activity, monitoring airglow, studying the distribution of unknown ultraviolet absorbing phenomenon at the cloud-tops, and making other science observations. It is derived in part by the *Mars Express* High Resolution Stereo Camera (HRSC) and the *Rosetta* Optical, Spectroscopic and Infrared Remote Imaging System (OSIRIS). The camera includes an FPGA to pre-process image data, reducing the amount transmitted to Earth.[13] The consortium of institutions responsible for the VMC includes the Max Planck Institute for Solar System Research, the Institute of Planetary Research at the German Aerospace Center and the Institute of Computer and Communication Network Engineering at Technische Universität Braunschweig.[14] It is not to be confused with Visual Monitoring Camera mounted on *Mars Express*, of which it is an evolution.[15][16]

**Magnetometer**

**MAG**: The magnetometer is designed to measure the strength of Venus's magnetic field and the direction of it as affected by the solar wind and Venus itself. It mapped the magnetosheath, magnetotail, ionosphere, and magnetic barrier in high resolution in three-dimensions, aid ASPERA-4 in the study of the interaction of the solar wind with the atmosphere of Venus, identify the boundaries between plasma regions, and carry planetary observations as well (such as the search for and characterization of Venus lightning). MAG is derived from the *Rosetta* lander's ROMAP instrument.

One measuring device is placed at the surface of the sonde, the identical second of the pair is placed a necessary distance off the body of the sonde by unfolding a 1 m long boom (carbon composite tube). Two redundant pyrotechnical cutters cut one loop of thin rope to free the power of metal springs. The driven knee lever rotates the boom perpendicularly outwards and latches at last. Only the use of a pair of sensors together with the rotation of the sonde allows to resolve the small natural magnetic field beneath the fields of the disturbing fields of the probe itself. The measurements took place already on the route from Earth to Venus.[17]

**Spectrometer**

**PFS**: The "Planetary Fourier Spectrometer" (PFS) operates in the infrared between the 0.9 μm and 45 μm wavelength range and is designed to perform vertical optical sounding of the Venus atmosphere. It performed global, long-term monitoring of the three-dimensional temperature field in the lower atmosphere (cloud level up to 100 kilometers). Furthermore it searched for minor atmospheric constituents that may be present, but had not yet been detected, analyzed atmospheric aerosols, and investigated surface to

atmosphere exchange processes. The design is based on a spectrometer on *Mars Express*, but modified for optimal performance for the *Venus Express* mission.

**SPICAV**: The "SPectroscopy for Investigation of Characteristics of the Atmosphere of Venus" (SPICAV) is an imaging spectrometer that was used for analyzing radiation in the infrared and ultraviolet wavelengths. It is derived from the *SPICAM* instrument flown on *Mars Express*. However, SPICAV has an additional channel known as **SOIR** (**S**olar **O**ccultation at **I**nfra**r**ed) that was used to observe the Sun through Venus's atmosphere in the infrared.

**VIRTIS**: The "**V**isible and **I**nfrared **T**hermal **I**maging **S**pectrometer" (VIRTIS) is an imaging spectrometer that observes in the near-ultraviolet, visible, and infrared parts of the electromagnetic spectrum. It analyzed all layers of the atmosphere, surface temperature and surface/atmosphere interaction phenomena.

**Radio science**

**VeRa**: **Ve**nus **Ra**dio Science is a radio sounding experiment that transmitted radio waves from the spacecraft and passed them through the atmosphere or reflected them off the surface. These radio waves were received by a ground station on Earth for analysis of the ionosphere, atmosphere and surface of Venus. It is derived from the Radio Science Investigation instrument flown on *Rosetta*.

## 1.37.3 Science

**Climate of Venus**

Starting out in the early planetary system with similar sizes and chemical compositions, the histories of Venus and Earth have diverged in spectacular fashion. It is hoped that the *Venus Express* mission data that was obtained can contribute not only to an in-depth understanding of how the Venusian atmosphere is structured, but also to an understanding of the changes that led to the current greenhouse atmospheric conditions. Such an understanding may contribute to the study of climate change on Earth.[18]

**Search for life on Earth**

*Venus Express* was also used to observe signs of life on Earth from Venus orbit. In images acquired by the probe, Earth was less than one pixel in size, which mimics observations of Earth-sized planets in other solar systems. These observations were then used to develop methods for habitability studies of exoplanets.[19]

### 1.37.4  Important events and discoveries

Important events for *Venus Express* include:

- 3 August 2005: *Venus Express* completed its final phase of testing at Astrium Intespace facility in Toulouse, France. It flew on an Antonov An-124 cargo aircraft via Moscow, before arriving at Baikonur on 7 August.

- 7 August 2005: *Venus Express* arrived at the airport of the Baikonur Cosmodrome.

- 16 August 2005: First flight verification test completed.

- 22 August 2005: Integrated System Test-3.

- 30 August 2005: Last Major System Test Successfully Started.

- 5 September 2005: Electrical Testing Successful.

- 21 September 2005: FRR (Fuelling Readiness Review) Ongoing.

- 12 October 2005: Mating to the Fregat upper stage completed.

- 21 October 2005: Contamination detected inside the fairing —launch on hold.

- 5 November 2005: Arrival at launch pad.

- 9 November 2005: Launch from Baikonur Cosmodrome at 03:33:34 UTC.

- 11 November 2005: First trajectory correction maneuver successfully performed.

- 17 February 2006: The main engine is fired successfully in a dress rehearsal for the arrival maneuver.[20]

- 24 February 2006: Second trajectory correction maneuver successfully performed.

- 29 March 2006: Third trajectory correction maneuver successfully performed - on target for 11 April orbit insertion.

- 7 April 2006: Command stack for orbit insertion maneuver is loaded on the spacecraft.

- 11 April 2006: The Venus Orbit Insertion (VOI) is completed successfully, according to the following timeline:[21]

    Period of this initial orbit is nine days.[6]

- 13 April 2006: First images of Venus from *Venus Express* released.

- 20 April 2006: Apoapsis Lowering Manoeuvre #1 performed. Orbital period is now 40 hours.

- 23 April 2006: Apoapsis Lowering Manoeuvre #2 performed. Orbital period is now approx 25 hours 43 minutes.

- 26 April 2006: Apoapsis Lowering Manoeuvre #3 is slight fix to previous ALM.

- 7 May 2006: *Venus Express* entered its target orbit at apoapsis at 13:31 UTC

- 14 December 2006: First temperature map of the southern hemisphere.

- 27 February 2007: ESA agrees to fund mission extension until May 2009.

- 19 September 2007: End of the nominal mission (500 Earth days) - Start of mission extension.

- 27 November 2007: The scientific journal *Nature* publishes a series of papers giving the initial findings. It finds evidence for past oceans. It confirms the presence of lightning on Venus and that it is more common on Venus than it is on Earth. It also reports the discovery that a huge double atmospheric vortex exists at the south pole of the planet.[22][23]

- 20 May 2008: The detection by the VIRTIS instrument on *Venus Express* of hydroxyl (OH) in the atmosphere of Venus is reported in the May 2008 issue of Astronomy and Astrophysics.[24]

- 4 February 2009: ESA agrees to fund mission extension until 31 December 2009.

- 7 October 2009: ESA agrees to fund the mission through 31 December 2012.

- 23 November 2010: ESA agrees to fund the mission through 31 December 2014.

- 25 August 2011: It is reported that a layer of ozone exists in the upper atmosphere of Venus.[25][26]

- 1 October 2012: It is reported that a cold layer where dry ice may precipate exists in the atmosphere of Venus.[27]

- 18 June —11 July 2014: Performs successful aerobraking experiment.[28][29]

- 28 November 2014: Mission control loses contact with *Venus Express*.[10]

- 3 December 2014: Intermittent contact established, spacecraft determined to likely be out of propellant.*[10]

- 16 December 2014: ESA declares the *Venus Express* mission over.*[4]

- 18 January 2015: Last detection of the spacecraft's X-band carrier signal.*[1]

### 1.37.5 See also

- Unmanned space mission

- Geosynchronous satellite

- List of planetary probes

- List of unmanned spacecraft by program

- Space exploration

- Space observatory

- Space probe

- Timeline of artificial satellites and space probes

- Timeline of planetary exploration

### 1.37.6 References

[1] Scuka, Daniel (23 January 2015). "Venus Express: The Last Shout". European Space Agency. Retrieved 26 January 2015.

[2] "Venturing into the upper atmosphere of Venus". European Space Agency. 11 November 2014. Retrieved 23 November 2014.

[3] "Operational Orbit". European Space Agency. 14 December 2012. Retrieved 23 November 2014.

[4] Bauer, Markus; Svedhem, Håkan; Williams, Adam; Martin, Patrick (16 December 2014). "Venus Express goes gently into the night". European Space Agency. Retrieved 22 December 2014.

[5] "Venus Express preliminary investigations bring encouraging news". ESA. 25 October 2005. Retrieved 9 May 2006.

[6] "Venus Express". *National Space Science Data Center*. Retrieved 22 December 2014.

[7] "Mission extensions approved for science missions". ESA. 16 October 2009.

[8] "Europe maintains its presence on the final frontier". ESA. 22 November 2010.

[9] "ESA science missions continue in overtime". ESA. 20 June 2013.

[10] "Venus Express anomaly". SpaceDaily. 8 December 2014. Retrieved 15 December 2014.

[11] Drake, Nadia (17 December 2014). "Out of Fuel, Venus Express Is Falling Gently to Its Death in Planet's Skies". *National Geographic*. Retrieved 22 December 2014.

[12] "The Venus Express mission camera". Max Planck Institute for Solar System Research.

[13] "Venus Monitoring Camera". Technical University at Brunswick.

[14] "The light and dark of Venus". ESA. 21 February 2008.

[15] Markiewicz, W. J.; Titov, D.; Fiethe, B.; Behnke, T.; Szemerey, I.; et al. "Venus Monitoring Camera for Venus Express" (PDF). Max Planck Institute for Solar System Research.

[16] Koeck, Ch.; Kemble, S.; Gautret, L.; Renard, P.; Faye, F. (October 2001). "Venus Express: Mission Definition Report" (PDF). European Space Agency. p. 17. ESA-SCI(2001)6.

[17] "IWF : VEX-MAG". Iwf.oewa.ac.at. Retrieved 15 December 2014.

[18] "Atmospheric Dynamics of Venus and Earth" (PDF). Lpi.usra.edu. Retrieved 15 December 2014.

[19] "Venus Express searching for life – on Earth". *European Space Agency*. Retrieved 15 December 2014.

[20] "Successful Venus Express main engine test". ESA. 17 February 2006. Retrieved 9 May 2006.

[21] "Venus orbit insertion". European Space Agency. 24 May 2007. Retrieved 26 January 2015.

[22] Various authors, Eric (November 2007). "European mission reports from Venus". *Nature* (450): 633–660. doi:10.1038/news.2007.297.

[23] "Venus offers Earth climate clues". *BBC News*. 28 November 2007. Retrieved 29 November 2007.

[24] "Venus Express Provides First Detection Of Hydroxyl In Atmosphere Of Venus". SpaceDaily. Retrieved 15 December 2014.

[25] Carpenter, Jennifer (7 October 2011). "Venus springs ozone layer surprise". *BBC News*.

[26] Montmessin, F.; Bertaux, J.-L.; Lefèvre, F.; Marcq, E.; Belyaev, D.; et al. (November 2011). "A layer of ozone detected in the nightside upper atmosphere of Venus". *Icarus* **216** (1): 82–85. Bibcode:2011Icar..216...82M. doi:10.1016/j.icarus.2011.08.010.

[27] "A curious cold layer in the atmosphere of Venus". *European Space Agency*. Retrieved 15 December 2014.

[28] Scuka, Daniel (16 May 2014). "Surfing an alien atmosphere". European Space Agency. Retrieved 23 November 2014.

[29] "Venus Express rises again". 11 July 2014. Retrieved 14 April 2015.

- Taylor, Fredric W. (November 2006). "The Planet Venus and the Venus Express Mission". *Planetary and Space Science* **54** (13-14): 1247–1248. Bibcode:2006P&SS...54.1247T. doi:10.1016/j.pss.2006.06.013.

- "Venus Express launch campaign starts". European Space Agency. 3 August 2005.

- "Venus Express Launch Campaign Journal". European Space Agency. Retrieved 16 August 2005.

- "Interactive 3D model of the Venus Express spacecraft". European Space Agency. Retrieved 5 September 2005.

- "Venus Express: Instruments". European Space Agency. Retrieved 14 September 2005.

### 1.37.7 Further reading

- Dambeck, Thorsten (2009). "The Blazing Hell Behind the Veil" (PDF). *MaxPlanckResearch* (4): 26–33. B56133.

### 1.37.8 External links

- *Venus Express* mission page by the European Space Agency

- *Venus Express* mission page by ESA Spacecraft Operations

- *Venus Express* profile by NASA's Solar System Exploration

# 1.38 Venus In Situ Explorer

The **Venus In-Situ Explorer** (**VISE**) is a mission proposed in 2003 by the Planetary Science Decadal Survey as a space probe designed to answer fundamental scientific questions by landing and performing experiments on Venus.[1] It is a candidate for NASA's New Frontiers program to be launched by 2022. The Principal Investigators are currently refining and promoting the mission concept.[2]

*An aerobot concept for Venus exploration*

### 1.38.1 Overview and capabilities

While on the surface, the Venus In-Situ Explorer will acquire and characterize a core sample of the surface to study pristine rock samples not weathered by the very harsh surface conditions of the planet. Also, the VISE will measure the composition and mineralogy of the surface.

### 1.38.2 Various exploration concepts

Several variant mission concepts for a Venus in-situ explorer have also been proposed, including both surface and atmospheric exploration:

- Other surface exploration concepts

- Aerobot concepts

- Venus airplane concepts

### 1.38.3 References

[1] Venus Exploration Analysis Group (VEXAG)

[2] Smith, B; Venkatapathy, E.; Wercinski, P.; Yount, B. (2013), "Venus In Situ Explorer Mission design using a mechanically deployed aerodynamic decelerator" (PDF), *2013 IEEE Aerospace ConferenceZ*, IEEE Explore, retrieved 2014-01-11

### 1.38.4 External links

- NASA Atmospheric Flight on Venus Landis, Geoffrey A., Colozza, Anthony, and LaMarre, Christopher M., International Astronautical Federation Congress 2002, paper IAC-02-Q.4.2.03, AIAA-2002-0819, AIAA0, No. 5

## 1.39 Venus orbiter mission

The **Venus orbiter mission** is a proposed orbiter to Venus by the Indian Space Research Organisation (ISRO) to study the atmosphere of Venus.[2][3] If funded, it would be launched in 2017 or 2018.

Jacques Blamont, an astrophysicist, has offered to provide the Indian Space Research Organisation with gigantic balloons carrying several instruments designed to deploy in and out of the extremely hot atmosphere of the planet after being unfettered from the orbiter.[4]

### 1.39.1 References

[1] "After Mars, Isro aims for Venus probe in 2-3 years". 9 June 2015.

[2] Ranosa, Ted (July 2015). "India Plans Mission To Venus Following Success Of Mars Orbiter". *Tech Times*. Retrieved 2015-10-13.

[3] Nowakowski, Tomasz (July 2015). "India eyes possible mission to Venus". *Spaceflight Insider*. Retrieved 2015-10-13.

[4] Srinivas Laxman (17 February 2012). "India planning Venus mission". Times of India. Retrieved 24 July 2012.

## 1.40 Venus Orbiting Imaging Radar

**Venus Orbiting Imaging Radar** (**VOIR**; also called **Venus Orbital Imaging Radar**) was a planned 1983 U.S. spacecraft mission to Venus that was primarily intended to use a microwave imaging radar to perform mapping of the Venusian surface. The goal was to map up to 50% of the planet's surface down to a resolution of 2 km with the eventual goal of targeting landers and atmospheric probes.[1] A 1978 study evaluated the potential use of synthetic aperture radar to achieve 200 meter resolution.[2] The spacecraft was to be launched from the Space Shuttle using a twin stage IUS in December 1984, and arrive in orbit May 1985. The mission was expected to last until November 1985.[3]

By 1981, the plan was for the spacecraft to launch in 1987 and to use aerobraking to circularize its orbit, whereupon it would be able to generate radar coverage of the entire planet over a period of 126 days. Data transmission rates were 1 Mbit per second, matching the imaging and recording speed. It would have two resolutions: mapping mode of 600 m per line-pair, then a high-resolution mode at 150 m per line-pair.[4]

The mission was cancelled in 1982 when it exceeded its budget limit. In 1983, it was replaced by a less ambitious mission called the Venus Radar Mapper, which was later renamed Magellan.[5]

### 1.40.1 References

[1] Rose, J. R.; Friedman, L. D. (1975), "Design for a Venus orbital imaging radar mission", *Journal of Spacecraft and Rockets* **12**: 106–112, Bibcode:1975JSpRo..12..106R, doi:10.2514/3.56954.

[2] Arens, W. E. (1978), "Real-time SAR image processing onboard a Venus orbiting spacecraft", *Proceedings of the 1978 Synthetic Aperture Radar Technology Conference 16*, Bibcode:1978sart.confR....A.

[3] Brown, C. D.; Frank, R. E. (1980), "Venus Orbital Imaging Radar (VOIR) mission: a further step in the exploration of Venus", *Acta Astronautica* **7**: 519–529, Bibcode:1980AcAau...7..519B.

[4] Leberl, F. W. (August 1981), "The Venus Orbital Imaging Radar / VOIR Mission", *The Solar System and its Exploration, Proceedings of the Alpach Summer School Conference*, p. 189, Bibcode:1981ESASP.164..189L.

[5] Evans, Ben (2012), *Tragedy and Triumph in Orbit: The Eighties and Early Nineties*, Springer Science & Business Media, p. 552.

## 1.41 VERITAS (spacecraft)

**VERITAS** (**Venus Emissivity, Radio Science, InSAR, Topography, and Spectroscopy**) is a proposed mission concept by NASA's Jet Propulsion Laboratory (JPL) to map with high resolution the surface of planet Venus. The combination of surface topography and image data would provide knowledge of Venus' tectonic and impact history, the timing and mechanisms of volcanic resurfacing, and the mantle processes responsible for them.

### 1.41.1 Overview

VERITAS was selected on 30 September 2015 as a semifinalist for Mission #13 of the Discovery Program.[1] The winner will be chosen around September 2016,[2] and must be ready to launch by the end of 2021.[3][4] Suzanne Smrekar of NASA's Jet Propulsion Laboratory (JPL) is the Principal Investigator, and JPL would manage the project.

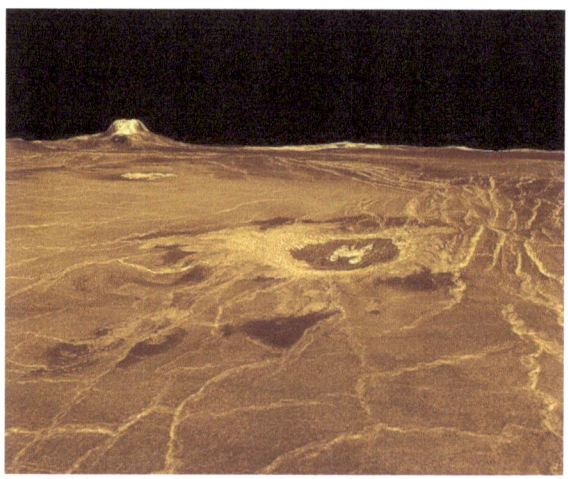

*This is an example of computer generated terrain of Venus, based on data from a orbiting radar imaging satellite. A new global data map would allow a comparison between the two to made*

*Size comparison of radar-mapped Venus surface and Earth*

## 1.41.2   Objectives

VERITAS would produce global, high resolution topography and imaging of Venus' surface and produce the first maps of deformation and global surface composition,[*][5] thermal emissivity, and gravity field.[*][6] It would also attempt to determine if Venus hosted ancient aqueous environments. Also, current data are highly suggestive of recent and active volcanism and this mission could determine if current volcanism is limited to mantle plume heads or is more widespread.[*][6]

High resolution imagery would be obtained by using an X band radar configured as a single pass interferometric synthetic aperture radar (InSAR)[*][7] coupled with a multispectral near-infrared (NIR) emissivity mapping capability. VERITAS would map surface topography with a spatial resolution of 250 m and 5 m vertical accuracy, and generate radar imagery with 30 m spatial resolution.[*][5]

**Goals** [*][7][*][8]

1. understand Venus' geologic evolution

2. determine what geologic processes are currently operating

3. find evidence for past or present water

## 1.41.3   Scientific payload

The primary mission goals, accomplished by seven objectives, require two instruments and a gravity science investigation over a 2-year orbital mission.[*][8]

- **VEM** (Venus Emissivity Mapper) maps surface emissivity using six spectral bands in five atmospheric windows that see through the clouds.[*][8] It would be provided by the German Aerospace Center (DLR)[*][9]

- **VISAR** (Venus Interferometric Synthetic Aperture Radar) generates a DEM (digital elevation model) with an accuracy of 250 m horizontal by 5 m height.[*][7]

Gravity science is carried out using the spacecraft's telecom system. The mission design also enables the opportunity to send a nanosat probe into the atmosphere of Venus, carrying a mass spectrometer to sample the noble gases and their isotopes.[*][8] For the NASA AO, this fulfills the option for a TDO (technology demonstration option)[*][10] Called *Cupid's arrow* it would pack a quadrupole ion trap mass spectrometer into nanosat atmospheric "skimmer".[*][11]

## 1.41.4   References

[1] Brown, Dwayne C.; Cantillo, Laurie (30 September 2015). "NASA Selects Investigations for Future Key Planetary Mission". *NASA News* (Washington, D.C.). Retrieved 2015-10-01.

[2] "Small Bodies Dominate NASA's Latest Discovery Competition". *SpaceNews.com*. July 7, 2015. Retrieved 2015-08-09.

[3] Clark, Stephen (24 February 2014). "NASA receives proposals for new planetary science mission". *Space Flight Now*. Retrieved 2015-02-25.

[4] Kane, Van (December 2, 2014). "Selecting the Next Creative Idea for Exploring the Solar System". *Planetary Society*. Retrieved 2015-02-10.

[5] Hensley, S.; Smrekar, S. E (2012). "VERITAS: A Mission Concept for the High Resolution Topographic Mapping and Imaging of Venus". *American Geophysical Union Fall Meeting 2012*. NASA. Retrieved 2015-10-01.

[6] Smrekar, S. E.; Elkins-Tanton, L. T; Hensley, S.; Campbell, B. A, B. A. (2014). *VERITAS: A mission to study the highest priority Decadal Survey questions for Venus.* American Geophysical Union - Fall Meeting 2014. NASA.

[7] Paller, M.; Figueroa, H.; Freeman, A.; et. al. (2015). *VISAR: A Next Generation Inteferometric Radar for Venus Exploration* (PDF). Venus Lab and Technology Workshop (2015). Universities Space Research Association.

[8] Freeman, A.; Smrekar, S. (9 June 2015). *VERITAS – a Discovery-class Venus surface geology and geophysics mission* (PDF). 11th Low Cost Planetary Missions Conference. Berlin, Germany.

[9] Helbert, J. (19 September 2013). *Observing the surface of Venus after VIRTIS on VEX: new concepts and laboratory work* (PDF). Infrared Remote Sensing and Instrumentation XXI. San Diego, USA.

[10]

[11]

### 1.41.5   See also

- DAVINCI (spacecraft) (Venus entry probe)

- Psyche (spacecraft) (Asteroid orbiter)

- Near-Earth Object Camera (NEOcam)

- Lucy (spacecraft) (Jupiter trojan tour)

- Magellan (spacecraft) (previous Venus radar mission)

### 1.41.6   External links

- Why We Explore - Venus (2006, NASA)

## 1.42   Vesta (spacecraft)

**Vesta** was a multiple-asteroid-flyby mission that the Soviet Union was planning in the 1980s.

The Vesta mission would have consisted of two identical probes (just like earlier Soviet Venus missions), to be launched in 1991. Similar to the Vega program, each spacecraft would deploy one or more landers or balloons into the Venusian atmosphere, and then proceed to its next target.

At Venus, a French satellite dedicated to asteroid flybys would be released.

It would return to us for an Earth swing-by, and then reach about 3-3.3 AUs from the Sun.

There they would fly by some smaller asteroids, and Vesta, if possible, with a small probe landing there.

The exact targets would depend on the launch date. In the initial 1985 study, 2700 possible trajectories were analyzed for a launch date in 1991/1992. Considering all constraints, about 12 candidate trajectories were selected. Of course, the two identical spacecraft could choose different trajectories and targets. These included 5 Astraea, 53 Kalypso, 187 Lamberta, 453 Tea, 1335 Demoulina and 1858 Lobachevskij, and comet Encke.

### 1.42.1   Vesta spacecraft design

Around 1985 Vesta was changed to be a Mars mission, with the asteroid-part unchanged. Detailed studies indicate each probe would have visited four small bodies, including asteroids belonging to different classes - providing a representative sample of the diversity of asteroids - and probably one or two comets as well.

Visiting at least one Apollo-Amor (Earth-nearing) asteroid was also given a preference.

Preliminary studies called for at least the following scientific instruments to be included:

- a wide angle camera (~6.5° field of view, 512x512 pixel CCD)

- a narrow angle camera (~0.5° field of view, 512x512 pixel CCD - 3.9 arcsec/pixel)

- a near-infrared spectrometer (measuring between 0.5-5 micrometers with lambda/delta lambda = 50, 5 arcminutes per pixel)

Possible further instrumentation:

- UV spectrometer (for imaging during a comet flyby)

- radar altimeter/radiometer

- a dust detector

- ion or neutral gas detector

Onboard memory would be about 240 Mbits. Images at closest approach (~500 km) could have a resolution of 10 m/pixel. Worst case downlink rate is 600 bit/second (if not using Deep Space Network (DSN)). The scientific payload is about 100 kg. The spacecraft has 750 kg dry mass, and carries 750 kg propellants, and possibly a 500 kg penetrator. 20 square meters of solar panels provide 350 Watts of power.

If DSN support could be obtained, Doppler tracking of the Vesta spacecraft's movement can be used to accurately determine the mass of the encountered bodies. In the other case, another possibility was considered: releasing a test mass, and observing its movement near the target asteroid.

The spacecraft's structure is derived from telecommunication satellites (INMARSAT), having the required mass, volume, and delta-v capabilities (3-axis stabilized, with a pointing platform with 2 axes of freedom for scientific instruments).

### 1.42.2  Trajectory

The Mars gravity assist constrain the possible trajectories. The asteroid penetrator also imposes limits on the speed of the approach of the target asteroid (less than 4 km/s).

Nevertheless, 3 possible trajectories were designed, with two Mars gravity assists.

A single Mars swing-by is also possible, but the double gravity assist increases the mass budget of the spacecraft by 30%, at the cost of an additional 1.8 year in travel time to the asteroid belt. The following trajectories are for the 1994 launch window. The size and type of each asteroid is also shown here:

Trajectory 1:

> launch from Earth
>
> Mars gravity assist
>
> flyby of 2335 James (a 10 km X-type asteroid) (an Amor-asteroid)
>
> Mars gravity assist
>
> 109 Felicitas (C-type, 76 km)
>
> 739 Mandeville (EMP(?) type, 110 km)
>
> 4 Vesta (V-type, or Vestoid. Has a diameter of 570 km) flyby with 3.5 km/s. A penetrator is released.

Total delta-v: 450 m/s

Trajectory 2:

> launch from Earth
>
> Mars gravity assist
>
> flyby of the 157P/Tritton short period comet
>
> Mars gravity assist
>
> 2087 Kochera(30 km?)
>
> 1 Ceres (flyby & releasing a penetrator)

Total delta-v: 1150 m/s

Trajectory 3:

> launch from Earth
>
> Mars gravity assist
>
> 1204 Renzia (10 km?) (an Amor-asteroid)
>
> Mars gravity assist
>
> 435 Ella (U type, 30 km)
>
> 46 Hestia (F type, 165 km)
>
> 135 Hertha (M type, 80 km)

Total delta-v: 350 m/s

In other studies 11 Parthenope, 19 Fortuna and 20 Massalia were also considered.

### 1.42.3  Cancellation

A combination of factors, probably including changing Franco-Soviet relations, the partial failure of the Phobos mission, financial troubles and the disbanding of the Soviet Union, prevented the project from advancing beyond the planning phase.

## 1.43  Zond 1

**Zond 1** was a member of the Soviet Zond program. It was the second Soviet research spacecraft to reach Venus, although communications had failed by that time. It carried a 90 cm spherical landing capsule, containing experiments for chemical analysis of the atmosphere, gamma-ray measurements of surface rocks, a photometer, temperature and pressure gauges, and a motion/rocking sensor in case it landed in water.

At least three previous Soviet planetary probes had been lost due to malfunctions of the ullage rockets (BOZ) on the Blok L stage, but an investigation found that the problem was easily resolved. The spacecraft, a *Venera 3MV−1*, was launched on April 2, 1964 from Tyuratam and this time the launch vehicle performed flawlessly. During the cruise phase, a slow leak from a cracked sensor window caused the electronics compartment to lose air pressure. This was a serious problem as Soviet electronics relied on vacuum tubes which would overheat without cooling air. An ill-timed command from ground control turned on its radio system while there was still a rarefied atmosphere inside, causing the electronics to short out by corona discharge. Since the mission could not succeed, it was unacceptable to give it a Venera designation, but as the probe had exited Earth orbit, it also could not be given the Kosmos cover name, so

*The Russian* Zond 1

Soviet authorities instead announced the probe as "Zond" and claimed that it was designed to "stress-test components in deep space". Chief Designer Sergei Korolev was upset at the failure of the mission and demanded higher quality control from the OKB-1 Bureau, including X-rays to test for pressure leaks.

By mid-April, the electronics in the main spacecraft had completely failed and all signal transmission ceased, however communication via the lander could still be performed and space radiation and atomic-hydrogen spectrometer measurements were received. The star trackers in the spacecraft were also used to align it for a course-correction burn, but the second one was off by 65 feet per second (20 meters per second). Also one of the star trackers failed, forcing ground controllers to place Zond 1 into a spin-stabilization mode. However, all communications had failed by May 14. It passed 100,000 km from Venus on July 14, 1964.

### 1.43.1 External links

- Astrolink description of spacecraft and payload

- NSSDC spacecraft info

## 1.44  3MV

The **3MV planetary probe** (short for 3rd generation Mars-Venus) is a designation for a common design used by early Soviet unmanned probes to Mars and Venus. It was an incremental improvement of earlier 2MV probes and was used for Zond 1, Zond 2 and Zond 3 missions to Mars as well as several Venera probes. It was standard practice of the Soviet space program to use standardized components as much as possible. All probes shared the same general characteristics and differed only in equipment necessary for specific missions. Each probe also incorporated improvements based on experience with earlier missions.

### 1.44.1  Design

*Zond 2 (interplanetary) part of 3MV family*

The probe consisted of three primary parts: The core of the stack was a pressurized compartment called the Orbital Compartment. This part housed the spacecraft's control electronics, radio transmitters and receivers, batteries, astro-orientation equipment, and so on. The compartment was pressurized to around 100 kPa and thermally controlled to simulate earth-like conditions, which removed the need for special electronic components that could reliably operate in extreme conditions (on Zond 1 the module depressurized in flight, severely damaging the probe's systems).

Mounted on the outside of the Orbital Compartment were

two solar panels which supplied power to the spacecraft. They were folded against the body of the probe during launch and were only deployed when the craft was already on its interplanetary trajectory. On the ends of each solar panel was a hemispherical radiator which radiated excess heat from the orbital compartment into space through a coolant loop.

Also mounted on the Orbital Compartment was a 2 m parabolical high-gain antenna, used for long-range communications. Depending on the mission, the probe also used other antennas (for example, for communication with probes on the planet's surface).

Below the Orbital Compartment was a second pressurized compartment called the Planetary Compartment. Depending on the mission the Planetary Compartment either housed scientific equipment for orbital observation of the planet or was designed to detach and land on the planet's surface.

Course correction capabilities were provided by a KDU 414 engine attached to the top of the Orbital Compartment. It provided a maximum thrust of around 2 kN and used UDMH and nitric acid as propellants. Attitude control was achieved by several small cold gas thrusters.

The whole stack was 3.6 m high and weighted around 1000 kg.

### 1.44.2   See also

- Soviet space program

- Zond program

- Venera

### 1.44.3   References

- "The Mystery of Zond 2" by Andrew Lepage, EJASA, April 1991, retrieved on January 21, 2012

- "The overview of the Russian launches toward Venus" by Anatoly Zak, July 2007, retrieved on December 2, 2007

- "The overview of the Russian launches toward Mars" by Anatoly Zak, August 2007, retrieved on December 2, 2007

## 1.45   Vega program

The **Vega program** was a series of Venus missions that also took advantage of the appearance of Comet Halley in 1986.

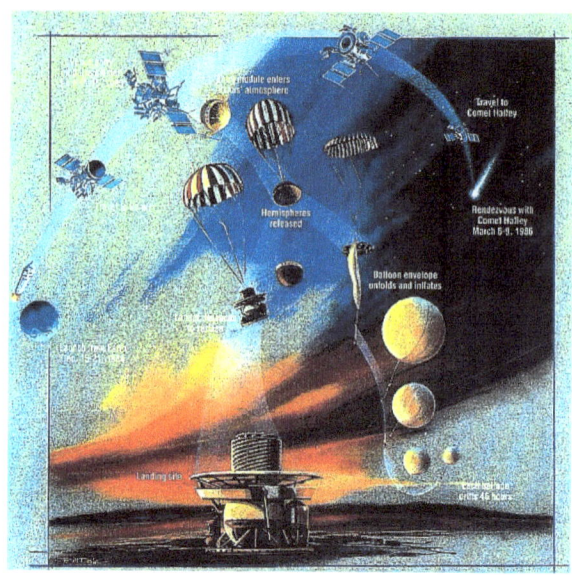

*Vega mission description*

Vega 1 and Vega 2 were unmanned spacecraft launched in a cooperative effort among the Soviet Union (who provided the spacecraft and launch vehicle) and Austria, Bulgaria, Hungary, the German Democratic Republic, Poland, Czechoslovakia, France, and the Federal Republic of Germany in December 1984. They had a two-part mission to investigate Venus and also flyby Halley's Comet.

The flyby of Halley's Comet had been a late mission change in the Venera program following on from the cancellation of the US Halley mission in 1981. A later Venera mission was cancelled and the Venus part of the Vega 1 mission was reduced. Because of this, the craft was designated Vega, a contraction of "Venera" and "Gallei" (Russian words for "Venus" and "Halley", respectively). The spacecraft design was based on the previous Venera 9 and Venera 10 missions.

The two spacecraft were launched on December 15 and 21, 1984, respectively. With their redesignated dual missions, the Vega probes became part of the Halley Armada, a group of space probes that studied Halley's Comet during its 1985/86 perihelion.

### 1.45.1   The Vega spacecraft

Vega 1 and 2 were identical sister ships. The spacecraft was a development of the earlier *Venera* craft. They were designed by Babakin Space Center and constructed as 5VK by Lavochkin at Khimki. The craft was powered by twin large solar panels and instruments included an antenna dish, cameras, spectrometer, infrared sounder, magnetometers (MISCHA), and plasma probes. The 4,920 kg craft was launched

*Vega solar system probe bus and landing apparatus (model)*

by a Proton 8K82K rocket from Baikonur Cosmodrome, Tyuratam, Kazakh SSR. Both Vega 1 and 2 were three-axis stabilized spacecraft. The spacecraft were equipped with a dual bumper shield for dust protection from Halley's comet.

## Bus Instruments

1. imaging system

2. infrared spectrometer

3. ultraviolet, visible, infrared imaging spectrometer

4. shield penetration detector

5. dust detectors

6. dust mass spectrometer

7. neutral gas mass spectrometer

8. APV-V plasma energy analyzer

9. energetic-particle analyzer

10. magnetometer

11. wave and plasma analyzers

## 1.45.2   The Venus mission

Vega 1 arrived at Venus on June 11, 1985 and Vega 2 on June 15, 1985, and each delivered a 1,500 kg, 240 cm diameter spherical descent unit. The units were released some days before each arrived at Venus and entered the atmosphere without active inclination changes. Each contained a lander and a balloon explorer.

## Descent craft

The landers were identical to that of the previous five *Venera* missions and were to study the atmosphere and surface, each had instruments to study temperature, pressure, a UV spectrometer, a water concentration meter, a gas-phase chromatograph, an X-ray spectrometer, a mass spectrometer and a surface sampling device.

The Vega 1 lander's surface experiments were inadvertently activated at 20 km from the surface by an especially hard wind jolt, and so failed to provide results. It landed at 7.5°N, 177.7°E.

The Vega 2 lander touched down at 03:00:50 UT on June 15, 1985 at 8.5° S, 164.5° E, in eastern Aphrodite Terra. The altitude of the touchdown site was 0.1 km above the planetary mean radius. The measured pressure at the landing site was 91 atm and the temperature was 736 K. The surface sample was found to be an anorthosite-troctolite. The lander transmitted data from the surface for 56 minutes.

## Payload

- Meteocomplex T,P sensors

- Sigma-3 gas chromatograph

- LSA particle size spectrometer

- IFP aerosol analyser

- VM-4 hygrometer

- ISAV-A nephelometer / scatterometer

- Malakhit-V mass spectrometer

- ISAV-S UV spectrometer

- GZU VB-02 drill + BDRP-AM25 soil X-ray fluorescence spectrometer

- GS-15-STsV gamma ray spectrometer

- PrOP-V penetrometer

- MSB small solar batteries

## Balloon

The two balloon aerobots were designed to float at 54 km from the surface, in the most active layer of the Venusian cloud system. The instrument pack had enough battery power for sixty hours of operation and measured temperature, pressure, wind speed and aerosol density. Both Vega-1 and Vega-2 balloons operated for more than 46 hrs from injection to the final transmission.[1]

The balloons were spherical superpressure types with a diameter of 3.54 metres (11.6 ft) and filled with helium. A gondola assembly weighing 6.9 kilograms (15.2 pounds) and 1.3 meters (4.26 ft) long was connected to the balloon envelope by a tether 13 metres (42.6 ft) long. Total mass of the entire assembly was 21 kilograms (46 pounds).

The top section of the gondola assembly was capped by a conical antenna 37 centimetres (14.6 inches) tall and 13 centimetres (5 $\frac{1}{8}$ inches) wide at the base. Beneath the antenna was a module containing the radio transmitter and system control electronics. The lower section of the gondola assembly carried the instrument payload and batteries.

The instruments consisted of:

- An arm carrying thin-film resistance thermometers and a velocity anemometer. The anemometer consisted of a free-spinning plastic propeller whose spin was measured by LED-photodetector optointerrupters.

- A module containing a PIN diode photodetector to measure light levels and a vibrating quartz beam pressure sensor.

- A package at the bottom carrying the batteries and a nephelometer to measure cloud density through light reflection.

The small low-power transmitter only allowed a data transmission rate of 2,048 bits per second, though the system performed data compression to squeeze more information through the narrow bandwidth. Nonetheless, the sampling rate for most of the instruments was only once every 75 seconds. The balloons were tracked by two networks of 20 radio telescopes in total back on Earth: the Soviet network, coordinated by the USSR Academy of Sciences and the international network, coordinated by CNES.

The balloons were dropped onto the planet's darkside and deployed at an altitude of about 50 kilometres (31 mi). They then floated upward a few kilometres to their equilibrium altitude. At this altitude, pressure and temperature conditions of Venus are similar to those of Earth, though the planet's winds moved at hurricane velocity and the carbon dioxide atmosphere is laced with sulfuric acid, along with smaller concentrations of hydrochloric and hydrofluoric acid.

The balloons moved swiftly across the night side of the planet into the light side, where their batteries finally ran down and contact was lost. Tracking indicated that the motion of the balloons included a surprising vertical component, revealing vertical motions of air masses that had not been detected by earlier probe missions.

*1985 USSR miniature sheet dedicated to the program, depicting Vega 1 spacecraft, Comet Halley and Intercosmos logo.*

### 1.45.3   The Halley mission

After their encounters, the Vega motherships were redirected by Venus's gravity to intercept Comet Halley.

Vega 1 made its closest approach on March 6, around 8,890 km from the nucleus, and Vega 2 made its closest approach on March 9 at 8,030 km. The data intensive examination of the comet covered only the three hours around closest approach. They were intended to measure the physical parameters of the nucleus, such as dimensions, shape, temperature and surface properties, as well as to study the structure and dynamics of the coma, the gas composition close to the nucleus, the dust particles' composition and mass distribution as functions of distance to the nucleus and the cometary-solar wind interaction.

In total Vega 1 and Vega 2 returned about 1,500 images of Comet Halley. Spacecraft operations were discontinued a few weeks after the Halley encounters.

The on-board TV system was created in international cooperation of the scientific and industrial facilities from the USSR, Hungary, France and Czechoslovakia. TV data were processed by international team, including the USSR, Hungary, France, GDR and USA scientists. The basic steps of data acquisition and preprocessing were performed in IKI using an image processing computer system based on a PDP-11/40 compatible host.

Vega 1 and 2 are currently in heliocentric orbits.

### 1.45.4 See also

- 5VK

  - Vega 1

  - Vega 2

- Venera program

- Pioneer Venus

### 1.45.5 References

[1] Preston; et al. (1986). "Determination of Venus Winds by Ground-Based Radio Tracking of the VEGA Balloons". *Science* **231** (4744): 1414–1416. Bibcode:1986Sci...231.1414P. doi:10.1126/science.231.4744.1414.

### 1.45.6 External links

- Vega mission images from the Space Research Institute (IKI)

- Raw data from Vega 1 and Vega 2 on board instruments

- Soviet Exploration of Venus

- HSAO/NASA Astrophysics Data System (ADS) – A tool for studying atmosphere dynamics on Venus

## 1.46 2V (V-69)

**2V (V-69)** may refer to satellites for which it was the manufacturer's designation:

- Venera 5

- Venera 6

## 1.47 Kosmos 21

**Kosmos 21** (Russian: Космос 21 meaning *Cosmos 21*) was a Soviet spacecraft with an unknown mission. This mission has been tentatively identified by NASA as a technology test of the Venera series space probes. It may have been an attempted Venus flyby, presumably similar to the later Kosmos 27 mission, or it may have been intended from the beginning to remain in geocentric orbit. In any case, the spacecraft never left Earth orbit after insertion by the SL-6/A-2-e launcher. The orbit decayed on November 14, three days after launch.

Cosmos 21 was launched at 06:23:35 UTC on 11 November 1963, atop a Molniya 8K78 carrier rocket flying from Site 1/5 at the Baikonur Cosmodrome. Its original development name before being given the Cosmos 21 denomination once it reached orbit was **3MV-1 No. 1**.[*][1]

Beginning in 1962, the name Kosmos was given to Soviet spacecraft which remained in Earth orbit, regardless of whether that was their intended final destination. The designation of this mission as an intended planetary probe is based on evidence from Soviet and non-Soviet sources and historical documents. Typically Soviet planetary missions were initially put into an Earth parking orbit as a launch platform with a rocket engine and attached probe. The probes were then launched toward their targets with an engine burn with a duration of roughly 4 minutes. If the engine misfired or the burn was not completed, the probes would be left in Earth orbit and given a Kosmos designation.

### 1.47.1 References

[1] McDowell, Jonathan. "Launch Log". Jonathan's Space Page. Retrieved 10 January 2011.

## 1.48 Kosmos 27

**Kosmos 27** (Russian: Космос 27 meaning *Cosmos 27*), also known as **Zond 3MV-1 No.3** was a space mission intended as a Venus flyby. The spacecraft was launched by a Molniya-M carrier rocket, however an upper stage malfunction resulted in the spacecraft failing to leave low Earth orbit.

Beginning in 1962, the name Kosmos was given to Soviet spacecraft which remained in Earth orbit, regardless of whether that was their intended final destination. The designation of this mission as an intended planetary probe is based on evidence from Soviet and non-Soviet sources and historical documents. Typically Soviet planetary missions were initially put into an Earth parking orbit as a launch platform with a rocket engine and attached probe. The probes were then launched toward their targets with an engine burn with a duration of roughly 4 minutes. If the engine misfired or the burn was not completed, the probes would be left in Earth orbit and given a Kosmos designation.

## 1.49   Kosmos 167

**Kosmos 167** (Russian: *Космос 167* meaning *Cosmos 167*), or **4V-1 No.311**, was a 1967 Soviet spacecraft intended to explore Venus. A 4V-1 spacecraft launched as part of the Venera programme, Kosmos 167 was intended to land on Venus, but never departed low Earth orbit due to a launch failure.

The 4V-1 No.311 spacecraft was the second of two 4V-1 vehicles built and operated by Lavochkin, following Venera 4.*[4]

A Molniya-M carrier rocket was used to launch 3MV-4 No.6. The launch occurred from Site 1/5 at the Baikonur Cosmodrome at 02:36:38 UTC on 17 June 1967.*[5] Due to a turbopump cooling problem, the rocket's Blok-L fourth stage failed to ignite, and as a result the spacecraft never departed its parking orbit.*[4] It was deployed into a low Earth orbit with a perigee of 187 kilometres (116 mi), an apogee of 262 kilometres (163 mi), and 51.8 degrees of inclination to the equator. The spacecraft was named *Kosmos 167*, part of a series typically used for military and experimental satellites in order to cover up the failure; had it departed Earth orbit it would have received the next designation in the *Venera* series, at the time Venera 5. Kosmos 167 was destroyed when it reentered the Earth's atmosphere on 25 June 1967.*[3]

### 1.49.1   References

[1] Krebs, Gunter. "Interplanetary Probes". *Gunter's Space Page*. Retrieved 11 April 2013.

[2] "Cosmos 167". US National Space Science Data Centre. Retrieved 11 April 2013.

[3] McDowell, Jonathan. "Satellite Catalog". *Jonathan's Space Page*. Retrieved 11 April 2013.

[4] Siddiqi, Asif A. (2002). "1967". *Deep Space Chronicle: A Chronology of Deep Space and Planetary Probes 1958-2000* (PDF). Monographs in Aerospace History, No. 24. NASA History Office. pp. 61–68.

[5] McDowell, Jonathan. "Launch Log". *Jonathan's Space Page*. Retrieved 11 April 2013.

## 1.50   Kosmos 482

**Kosmos 482** (Russian: *Космос 482* meaning *Cosmos 482*), launched March 31, 1972 at 04:02:33 UTC, was an attempted Venus probe which failed to escape low Earth orbit.

Beginning in 1962, the name Kosmos was given to Soviet spacecraft which remained in Earth orbit, regardless of whether that was their intended final destination. The designation of this mission as an intended planetary probe is based on evidence from Soviet and non-Soviet sources and historical documents. Typically Soviet planetary missions were initially put into an Earth parking orbit as a launch platform with a rocket engine and attached probe. The probes were then launched toward their targets with an engine burn with a duration of roughly 4 minutes. If the engine misfired or the burn was not completed, the probes would be left in Earth orbit and given a Kosmos designation.

Kosmos 482 was launched by a Molniya 8K78M booster on March 31, 1972, 4 days after the Venera 8 atmospheric probe and may have been similar in design and mission plan. After achieving an Earth parking orbit, the spacecraft made an apparent attempt to launch into a Venus transfer trajectory. It separated into four pieces, two of which remained in low Earth orbit and decayed within 48 hours into south New Zealand (known as the *Ashburton balls incident*), and two pieces (presumably the payload and detached engine unit) went into a higher 210 x 9800 km orbit. An incorrectly set timer caused the Blok L stage to cut off prematurely, preventing the probe from escaping Earth orbit.

At 1:00 AN on April 3, 1972, four red-hot 13.6 kg titanium alloy balls landed within a 16 km radius of each other just outside Ashburton, New Zealand.*[1] The 38 cm-diameter spheres scorched holes in crops and made deep indentations in the soil, but no one was injured. A similarly shaped object was discovered near Eiffelton, New Zealand, in 1978.

Space law required that the space junk be returned to its national owner, but the Soviets denied knowledge or ownership of the satellite. Ownership therefore fell to the farmer upon whose property the satellite fell. Kosmos 482 was thoroughly analyzed by New Zealand scientists which determined that they were Soviet in origin because of manufacturing marks and the high-tech welding of the titanium. The scientists concluded that they were probably gas pressure vessels of a kind used in the launching rocket for a satellite or space vehicle and had decayed in the atmosphere.

NSSDC ID: 1972-023A

NORAD ID: 05919

### 1.50.1   See also

- Russian space program
- Cosmos (satellite)

### 1.50.2   References

[1] "New light on mysterious space balls". *New Zealand Herald*. 2002-08-24. Retrieved 2006-10-08.

### 1.50.3  External links

- Space.com: Aussies, Kiwis Take Mir Deorbit in Stride 02:11 pm ET February 20, 2001

- Wired Magazine: Awaiting Mir's Crash Down Under 02:00 AM Feb, 19, 2001

## 1.51  Kosmos 96

**Kosmos 96** (Russian: *Космос 96* meaning *Cosmos 96*), or **3MV-4 No.6**, was a Soviet spacecraft intended to explore Venus. A 3MV-4 spacecraft launched as part of the Venera programme, Kosmos 96 was to have made a flyby of Venus, however due to a launch failure it did not depart low Earth orbit.

The 3MV-4 No.6 spacecraft was originally built for a mission to Mars, with launch scheduled for late 1964. After it was not launched by the end of its launch window, the spacecraft was repurposed, along with two other spacecraft which were launched as Venera 2 and Venera 3, to explore Venus.[4]

A Molniya carrier rocket was used to launch 3MV-4 No.6. The launch occurred from Site 31/6 at the Baikonur Cosmodrome at 03:21 UTC on 23 November 1965.[2] Late in third stage flight, a fuel line ruptured, causing one of the engine's combustion chambers to explode. The rocket tumbled out of control, and as a result the fourth stage, a Blok-L, failed to ignite.[4] The spacecraft was deployed into a low Earth orbit with a perigee of 209 kilometres (130 mi), an apogee of 261 kilometres (162 mi), and 51.9 degrees of inclination to the equator. The spacecraft was named *Kosmos 96*, part of a series typically used for military and experimental satellites in order to cover up the failure. Had it departed Earth's orbit, it would have received the next designation in the *Venera* series, at the time Venera 4.

Kosmos 96 was destroyed when it reentered the Earth's atmosphere on 9 December 1965.[3] Its reentry has been suggested as a possible explanation of UFO sightings over the United States and Canada, centred on Kecksburg, Pennsylvania; however analysis found the spacecraft probably reentered several hours before the sightings.[5]

### 1.51.1  References

[1] Krebs, Gunter. "Interplanetary Probes". *Gunter's Space Page*. Retrieved 11 April 2013.

[2] McDowell, Jonathan. "Launch Log". *Jonathan's Space Page*. Retrieved 11 April 2013.

[3] McDowell, Jonathan. "Satellite Catalog". *Jonathan's Space Page*. Retrieved 11 April 2013.

[4] Siddiqi, Asif A. (2002). "1965". *Deep Space Chronicle: A Chronology of Deep Space and Planetary Probes 1958-2000* (PDF). Monographs in Aerospace History, No. 24. NASA History Office. pp. 47–52.

[5] "Cosmos 96". US National Space Science Data Centre. Retrieved 11 April 2013.

## 1.52  Venera 2

**Venera 2** (Russian: Венера−2 meaning *Venus 2*), also known as **3MV-4 No.4** was a Soviet spacecraft intended to explore Venus. A 3MV-4 spacecraft launched as part of the Venera programme, it failed to return data after flying past Venus.

Venera 2 was launched by a Molniya carrier rocket, flying from Site 31/6 at the Baikonur Cosmodrome.[2] The launch occurred at 05:02 UTC on 12 November 1965, with the first three stages placing the spacecraft and Blok-L upper stage into a low Earth parking orbit before the Blok-L fired to propel Venera 2 into heliocentric orbit bound for Venus.

The Venera 2 spacecraft was equipped with cameras, as well as a magnetometer, solar and cosmic x-ray detectors, piezoelectric detectors, ion traps, a Geiger counter and receivers to measure cosmic radio emissions.[3] The spacecraft made its closest approach to Venus at 02:52 UTC on 27 February 1966, at a distance of 24,000 kilometres (15,000 mi).[4]

During the flyby, all of Venera 2's instruments were activated, requiring that radio contact with the spacecraft be suspended. The probe was to have stored data using onboard recorders, and then transmitted it to Earth once contact was restored. Following the flyby the spacecraft failed to reestablish communications with the ground. It was declared lost on 4 March.[3] An investigation into the failure determined that the spacecraft had overheated due to a radiator malfunction.[3]

### 1.52.1  References

[1] Krebs, Gunter. "Interplanetary Probes". *Gunter's Space Page*. Retrieved 11 April 2013.

[2] McDowell, Jonathan. "Launch Log". *Jonathan's Space Page*. Retrieved 11 April 2013.

[3] Siddiqi, Asif A. (2002). "1965". *Deep Space Chronicle: A Chronology of Deep Space and Planetary Probes 1958-2000* (PDF). Monographs in Aerospace History, No. 24. NASA History Office. pp. 47–52.

[4]  "Venera 2" . US National Space Science Data Centre. Retrieved 11 April 2013.

## 1.53   Venera 3

**Venera 3** (Russian: Венера–3 meaning *Venus 3*) (Manufacturer's Designation: 3MV-3) was a Venera program space probe that was built and launched by the Soviet Union to explore the surface of Venus. It was launched on 16 November 1965 at 04:19 UTC from Baikonur, Kazakhstan.

### 1.53.1   Mission

The mission of this spacecraft was to land on the Venusian surface. The entry body contained a radio communication system, scientific instruments, electrical power sources, and medallions bearing the Coat of Arms of the Soviet Union.

The probe possibly crash-landed on Venus on 1 March 1966, making Venera 3 the first spacecraft to impact on the surface of another planet. However, its communications systems failed before it reached the planet.[1][2]

### 1.53.2   See also

- 1965 in spaceflight

- Timeline of planetary exploration

### 1.53.3   References

[1]  David Leverington (2000). *New cosmic horizons.* Cambridge University Press. p. 74. ISBN 0-521-65833-0.

[2]  "Venera 3" . NASA.

## 1.54   Venera 4

**Venera 4** (Russian: Венера–4 meaning *Venus 4*), also designated **4V-1 No.310**[1] was a probe in the Soviet Venera program for the exploration of Venus. It was the first successful probe to perform in-place analysis of the environment of another planet. It may also have been the first probe to land on another planet, with the fate of its predecessor Venera 3 being unclear.[4] Venera 4 provided the first chemical analysis of the Venusian atmosphere, showing it to be primarily carbon dioxide with a few percent of nitrogen and below one percent of oxygen and water vapors. The station detected a weak magnetic field and no radiation field. The outer atmospheric layer contained very little hydrogen and no atomic oxygen. The probe sent the first direct measurements proving that Venus was extremely hot, that its atmosphere was far denser than expected, and that it had lost most of its water long ago.

### 1.54.1   Design

*Venera 4 hub*

The main hub of Venera 4 stood 3.5 metres (11 ft) high, its solar panels spanned 4 metres (13 ft) and had an area of 2.5 square metres (27 sq ft). The hub included a 2 meter long magnetometer, an ion detector, a cosmic ray detector and an ultraviolet spectrometer capable of detecting hydrogen and oxygen gases. The devices were intended to operate until entry into the Venusian atmosphere. At that juncture, the station was designed to release the probe capsule and disintegrate. The rear part of the hub contained a liquid-fuel thruster capable of correcting the flight course. The flight program was planned to include two significant course corrections, for which purpose the station could receive and execute up to 127 different commands sent from the Earth.[5]

The front part of the hub contained a nearly spherical landing capsule 1 meter in diameter and weighing 383 kg. Compared to previous (failed) Venera probes, the capsule contained an improved heat shield which could withstand temperatures up to 11,000 °C (19,800 °F). Instead of the previous liquid-based cooling design, a simpler and more reliable gas system was installed.[2] The durability of the capsule was checked by exposing it to high temperatures, pressures and accelerations using three unique testing installations. The heat resistance was checked in a high-temperature vacuum system emulating the upper layers of the atmosphere.[6] The capsule was also pressurized up to 25 atmospheres. (The surface pressure on Venus was unknown at the time. Estimates ranged from a few to hundreds of atmospheres).[7] Finally, it was subjected to accelerations of up to 450 G in a centrifuge. The centrifuge test caused cracking of electronic components and cable brackets, which were all replaced shortly before launch. The timing for launch was rather tight, so as not to miss the "launch window" – the days of the year when the path to the destination planet from Earth is energetically least demanding.

The capsule could float in case of a water landing. Considering the possibility of such a landing, its designers made the lock of the capsule using sugar;[5][6][8] it was meant to dissolve in liquid water, releasing the transmitter antennas. The capsule contained a newly developed vibration-damping system and its parachute could resist temperatures up to 450 °C.[6]

The capsule contained an altimeter, thermal control, a parachute and equipment for making atmospheric measurements. The latter included a thermometer, barometer, hydrometer, altimeter and a set of gas analysis instruments. The data were sent by two transmitters at a frequency of 922 MHz and a rate of 1 bit/s; the measurements were sent every 48 seconds. The transmitters were activated by the parachute deployment as soon as the outside pressure reached 0.6 standard atmospheres (61 kPa), which was thought to occur at the altitude about 26 kilometres (16 mi) above the surface of the planet. The signals were received by several stations, including the Jodrell Bank Observatory.[2][5]

The capsule was equipped with a rechargeable battery with a capacity sufficient for 100 minutes of powering the measurement and transmitter systems. To avoid becoming discharged during the flight to Venus, the battery was kept charged using the solar panels of the hub. Before the launch, the entire Venera 4 station was sterilized to prevent possible biological contamination of Venus.[2]

### 1.54.2  The mission

Two nominally identical 4V-1 probes were launched in June 1967. The first probe, Venera 4, was launched on 12 June by a Molniya-M carrier rocket flying from the Baikonur Cosmodrome.[1] A course correction was performed on 29 July when it was 12 million km away from Earth; otherwise the probe would have missed Venus. Although two such corrections had been planned, the first one was accurate enough and therefore the second correction was canceled. On October 18, 1967, the spacecraft entered the Venusian atmosphere with an estimated landing place near 19°N 38°E / 19°N 38°E.[5] The second probe, Kosmos 167, was launched on 17 June but failed to depart low Earth orbit.[9]

During entry into the Venusian atmosphere, the heat shield temperature rose to 11,000 °C (19,800 °F) and at one point the cabin deceleration reached 300 G.[10] The descent lasted 93 minutes. The capsule deployed its parachute at an altitude of about 52 kilometres (32 mi), and started sending data on pressure, temperature and gas composition back to Earth. The temperature control kept the inside of the capsule at −8 °C (18 °F). The temperature at 52 km was recorded as 33 °C (91 °F), and the pressure as less than 1 standard atmosphere (100 kPa). At the end of the 26-km descent, the temperature reached 262 °C (504 °F) and pressure increased to 22 standard atmospheres (2,200 kPa), and the signal transmission terminated. The atmospheric composition was measured as 90–93% carbon dioxide, 0.4–0.8% oxygen, 7% nitrogen and 0.1–1.6% water vapor.[5]

Malfunction of the altimeter resulted in the value of initial altitude (deployment of the capsule's parachute and start of the measurements) being transmitted as 26 kilometres (16 mi). Therefore, some Earth observers interpreted the descent as having continued to the surface of Venus, which was quickly dismissed as inconsistent with other data. In particular, the pressure readings by the capsule were much too low for the Venusian surface.[2][6]

### 1.54.3  Achievements

For the first time, *in situ* analysis of the atmosphere of another planet was performed and the data sent back to Earth; the analysis included chemical composition, temperature and pressure. The measured ratio of carbon dioxide to nitrogen of about 13 corrected the previous estimates so much (an inverse ratio was expected in some quarters) that some scientists contested the observations. The main station detected no radiation belts; relative to Earth, the measured magnetic field was 3000 times weaker, and the hydrogen corona was 1000 times less dense. No atomic oxygen was detected. All the data suggested that water, if present, had leaked from the planet long before. This conclusion was unexpected considering the thick Venusian clouds. Because of the negligible humidity, the sugar lock system, employed on Venera 4 in case of a water landing, was abandoned in subsequent Venus probes.[2][5]

The mission was considered a complete success, especially given several previous failures of Venera probes.[*][2] Although the Venera 4 design did allow for data transmission after landing, the Venera 3–6 probes were not built to withstand the pressures at the Venusian surface. The first successful landing on Venus was achieved by Venera 7 in 1970.

### 1.54.4  References

[1] McDowell, Jonathan.  "Launch Log" .  *Jonathan's Space Page*. Retrieved 11 April 2013.

[2]  "Venera 4 (in Russian)". Retrieved 2009-07-07.

[3] "Spacecraft - Details". National Space Science Data Center. Retrieved 2013-11-05.

[4] David Leverington (2000).  *New cosmic horizons*.  Cambridge University Press. pp. 74–74. ISBN 0-521-65833-0.

[5] Brian Harvey (2007).  *Russian planetary exploration*. Springer. pp. 98–101. ISBN 0-387-46343-7.

[6] Paolo Ulivi, David Michael Harland (2007).  *Robotic Exploration of the Solar System: The golden age 1957–1982*. Springer. pp. 55–56. ISBN 0-387-49326-3.

[7] Vakhnin, V. M. (1968).  "A Review of the Venera 4 Flight and Its Scientific Program" .  *J. Atmos. Sci.*  **25**: 533–534.    Bibcode:1968JAtS...25..533V.  doi:10.1175/1520-0469(1968)025<0533:AROTVF>2.0.CO;2.

[8] Photo of the lock.  novosti-kosmonavtiki.ru, 18 February 2005.

[9]  "Cosmos 167" .  *National Space Science Data Center*. National Aeronautics and Space Administration.  Retrieved 2014-02-13.

[10] Paolo Ulivi, David Michael Harland (2007).  *Robotic Exploration of the Solar System: The golden age 1957–1982*. Springer. p. 63. ISBN 0-387-49326-3.

## 1.55    Venera 5

**Venera 5** (Russian: Венера−5 meaning *Venus 5*) (manufacturer's designation: 2V (V-69)) was a probe in the Soviet space program *Venera* for the exploration of Venus.

Venera 5 was launched towards Venus to obtain atmospheric data. The spacecraft was very similar to Venera 4 although it was of a stronger design. The launch was conducted using a Molniya-M rocket, flying from the Baikonur Cosmodrome.

When the atmosphere of Venus was approached, a capsule weighing 405 kg and containing scientific instruments was

jettisoned from the main spacecraft.  During satellite descent towards the surface of Venus, a parachute opened to slow the rate of descent. For 53 minutes on May 16, 1969, while the capsule was suspended from the parachute, data from the Venusian atmosphere were returned.[*][1] It landed at 3°S 18°E / 3°S 18°E. The spacecraft also carried a medallion bearing the State Coat of Arms of the USSR and a bas-relief of V. I. Lenin to the night side of Venus.

Given the results from Venera 4, the Venera 5 and Venera 6 landers contained new chemical analysis experiments tuned to provide more precise measurements of the atmosphere's components.  Knowing the atmosphere was extremely dense, the parachutes were also made smaller so the capsule would reach its full crush depth before running out of power (as Venera 4 had done).

### 1.55.1    References

[1] Anne Marie Helmenstine,  "This Day in Science History - May 16 - Venera 5 'Landing'", About.com

### 1.55.2    External links

- Astronautix.com

## 1.56    Venera 6

**Venera 6** (Russian: Венера−6 meaning *Venus 6*), manufacturer's designation:  **4V-1 No.331**, was a Soviet spacecraft, launched towards Venus to obtain atmospheric data. It had an on-orbit dry mass of 1,130 kilograms (2,490 lb).

The spacecraft was very similar to Venera 4 although it was of a stronger design. When the atmosphere of Venus was approached, a capsule with a mass of 405 kilograms (893 lb) was jettisoned from the main spacecraft. This capsule contained scientific instruments.

During descent towards the surface of Venus, a parachute opened to slow the rate of descent. For 51 min on May 17, 1969, while the capsule was suspended from the parachute, data from the Venusian atmosphere were returned. It landed at 5°S 23°E / 5°S 23°E.

The spacecraft also carried a medallion bearing the State Coat of Arms of the U.S.S.R. and a bas-relief of V. I. Lenin to the night side of Venus.

Given the results from Venera 4, the Venera 5 and Venera 6 landers contained new chemical analysis experiments tuned to provide more precise measurements of the atmosphere's components.  Knowing the atmosphere was extremely dense, the parachutes were also made smaller so the capsule would reach its full crush depth before running out of power (as Venera-4 had done).

### 1.56.1 External links

- Astronautix.com

## 1.57 Venera 2MV-1 No.1

**Venera 2MV-1 No.1**,[*][1][*][2] also known as **Sputnik 19** in the West, was a Soviet spacecraft, which was launched in 1962 as part of the Venera programme, and was intended to become the first spacecraft to land on Venus.[*][3] Due to a problem with its upper stage it failed to leave low Earth orbit, and reentered the atmosphere a few days later.[*][4] It was the first of two Venera 2MV-1 spacecraft, both of which failed to leave Earth orbit.[*][2]

Venera 2MV-1 No.1 was launched at 02:18:45 UTC on 25 August 1962, atop a Molniya 8K78 carrier rocket flying from Site 1/5 at the Baikonur Cosmodrome.[*][1] The first three stages of the rocket operated nominally, injecting the fourth stage and payload into a low Earth orbit. The fourth stage then coasted until one hour and fifty seconds after launch, when it fired its ullage motors in preparation for ignition. One of the ullage motors failed to fire, and when the main engine ignited for a four-minute burn to place the spacecraft into heliocentric orbit, the stage began to tumble out of control. Forty-five seconds later, its engine cut off, leaving the spacecraft stranded in Earth orbit.[*][5] It reentered the atmosphere on 28 August 1962, three days after it had been launched.[*][6]

The designations Sputnik 23, and later Sputnik 19 was used by the United States Naval Space Command to identify the spacecraft in its Satellite Situation Summary documents, since the Soviet Union did not release the internal designations of its spacecraft at that time, and had not assigned it an official name due to its failure to depart geocentric orbit.[*][3][*][7][*][8]

### 1.57.1 References

[1] McDowell, Jonathan. "Launch Log". Jonathan's Space Page. Retrieved 28 July 2010.

[2] Krebs, Gunter. "Venera (2a), (2b) (2MV-1 #1, 2)". Gunter's Space Page. Retrieved 28 July 2010.

[3] Zak, Anatoly. "Russia's unmanned missions to Venus". RussianSpaecWeb. Retrieved 28 July 2010.

[4] Wade, Mark. "Venera". Encyclopedia Astronautica. Retrieved 28 July 2010.

[5] Wade, Mark. "Soyuz". Encyclopedia Astronautica. Retrieved 28 July 2010.

[6] McDowell, Jonathan. "Satellite Catalog". Jonathan's Space Page. Retrieved 28 July 2010.

[7] Robbins, Stuart J. (11 January 2006). "Soviet Craft - Sputnik". Journey Through The Galaxy. Retrieved 28 July 2010.

[8] "Sputnik 19". NASA NSSDC. Retrieved 28 July 2010.

## 1.58 Venera 4V-2

*Venera 15 / 16*

**Venera 4V-2** (Russian: Венера 4В−2) was a series of two identical spacecraft sent to Venus by the Soviet Union, consisting of Venera 15 and Venera 16.[*][1] Both unmanned orbiters were to map the surface of Venus using high resolution imaging systems. The spacecraft were identical and based on modifications to the earlier Venera space probes.

### 1.58.1 Mission profile

Venera 15 was launched on June 2, 1983 at 02:38:39 UTC, and Venera 16 on June 7, 1983 at 02:32:00 UTC. Venera

15 and Venera 16 both reached Venus' orbit (on October 10, 1983 and October 14, 1983 respectively).

The two spacecraft were inserted into Venus orbit a day apart with their orbital planes shifted by an angle of approximately 4° relative to one another. This made it possible to reimage an area if necessary. Each spacecraft was in a nearly polar orbit with a periapsis ~1000 km, at 62°N latitude, and apoapsis ~65000 km, with an inclination ~90°, the orbital period being ~24 hours.

In June 1984, Venus was at superior conjunction and passed behind the Sun as seen from Earth. No transmissions were possible, so the orbit of Venera 16 was rotated back 20° at this time to map the areas missed during this period.

Together, the two spacecraft imaged the area from the north pole down to about 30°N latitude (i.e. approx. 25% of Venus surface) over the 8 months of mapping operations.

## 1.58.2   Spacecraft structure

The Venera 15 and 16 spacecraft were identical and were based on modifications to the orbiter portions of the Venera 9 and Venera 14 probes. Each spacecraft consisted of a 5 m long cylinder with a 0.6 m diameter, 1.4 m tall parabolic dish antenna for the synthetic aperture radar (SAR) at one end. A 1-meter diameter parabolic dish antenna for the radio altimeter was also located at this end. The electrical axis of the radio altimeter antenna was lined up with the axis of the cylinder. The electrical axis of the SAR deviated from the spacecraft axis by 10 degrees. During imaging, the radio altimeter would be lined up with the center of the planet (local vertical) and the SAR would be looking off to the side at 10 degrees. A bulge at the opposite end of the cylinder held fuel tanks and propulsion units. Two square solar arrays extended like wings from the sides of the cylinder. A 2.6 m radio dish antenna for communications was also attached to the side of the cylinder. The spacecraft each massed 4000 kg.

Both Venera 15 and 16 were equipped with a Synthetic Aperture Radar (SAR). A radar was necessary in this mission because nothing else would be able to penetrate the dense clouds of Venus. The probes were equipped with on board computers that saved the images until the entire image was complete.

## 1.58.3   See also

- Venera 15

- Venera 16

## 1.58.4   References

[1]  Wade, Mark. "Venera 4V-2". *Encyclopedia Astronautica*. Retrieved 14 January 2011.

## 1.58.5   External links

- The Soviet Exploration of Venus

- Catalog of Soviet Venus images

- Venera 16 (National Space Science Data Center, NASA)

# Chapter 2

# Text and image sources, contributors, and licenses

## 2.1 Text

- **Akatsuki (spacecraft)** *Source:* https://en.wikipedia.org/wiki/Akatsuki_(spacecraft)?oldid=694575375 *Contributors:* Vicki Rosenzweig, Molinari, Twang, Xanzzibar, Fukumoto, Andycjp, Zeimusu, O'Dea, CALR, JTN, Rich Farmbrough, Bender235, Russ3Z, Hooperbloob, A2Kafir, Grutness, Bricktop, Tabletop, Eyreland, とある白い猫, Emerson7, Pmj, Rjwilmsi, Vegaswikian, FlaBot, Nihiltres, Gurch, Kolbasz, DVdm, The Rambling Man, YurikBot, Baumi, RadioFan, Hydrargyrum, NawlinWiki, Nick, Voidxor, Reyk, Mhenriday, SmackBot, Nickst, Yamaguchi 先生, George Ho, WDGraham, VJDocherty, Chlewbot, Aces lead, Ohconfucius, Freewol, JorisvS, RandomCritic, Xiaphias, Novangelis, Swotboy2000, Joseph Solis in Australia, Negadrive~enwiki, ThreeBlindMice, Ruslik0, N2e, Cydebot, PamD, Thijs!bot, Headbomb, Ricnun, JAnDbot, BatteryIncluded, Torchiest, GRAND OUTCAST, CommonsDelinker, Woodwhite, Rod57, Ohms law, Tambora1815, STBotD, ACSE, VolkovBot, Rei-bot, JhsBot, Weetjesman, Djmckee1, Petergans, Mycomp, Oda Mari, Lightmouse, Drsteevo, RSStockdale, Dabomb87, MBK004, Savacek, TypoBoy, Ktr101, Alexbot, PixelBot, SounderBruce, Addbot, Roentgenium111, LaaknorBot, Lightbot, Luckas-bot, Shinkansen Fan, AnomieBOT, ArthurBot, Xqbot, Skore de, GrouchoBot, Trurle, RibotBOT, Trafford09, Fotaun, FrescoBot, 117Avenue, DrilBot, Phoenix7777, TedderBot, IJBall, IVAN3MAN, FoxBot, Trappist the monk, RjwilmsiBot, DASHBot, John of Reading, Wikitanvir Bot, Ozric14, GoingBatty, Infinitjest, Yiosie2356, H3llBot, SporkBot, ChiZeroOne, Mjbmrbot, ClueBot NG, Violettsureme, Frietjes, Bibcode Bot, BG19bot, Kendall-K1, B2322858, Joydeep, Dexbot, Sspacce, Athomeinkobe, Dschslava, Coladar, Stilgar27, Exoplanetaryscience, Mfb, Monkbot, Kjerish, Quivico, LL221W, DN-boards1, JadeDragon4001, Ceannlann gorm, CaptainBrant, Firebrace and Anonymous: 53

- **DAVINCI (spacecraft)** *Source:* https://en.wikipedia.org/wiki/DAVINCI_(spacecraft)?oldid=691836416 *Contributors:* BatteryIncluded, Emeraude, Fotaun, John of Reading and Ninney

- **European Venus Explorer** *Source:* https://en.wikipedia.org/wiki/European_Venus_Explorer?oldid=677950857 *Contributors:* Cyclopia, Wjfox2005, Nickst, WDGraham, ThreeBlindMice, Cydebot, Afterthewar~enwiki, Danmichaelo, Addbot, Lightbot, Luckas-bot, Yobot, AnomieBOT, Fotaun, EmausBot, ZéroBot, Yiosie2356, ChiZeroOne, Jjbernardiscool, Isambard Kingdom, DN-boards1 and Anonymous: 6

- **High Altitude Venus Operational Concept** *Source:* https://en.wikipedia.org/wiki/High_Altitude_Venus_Operational_Concept?oldid=693239444 *Contributors:* Stevenjgarner, JorisvS, BatteryIncluded, Fotaun, I dream of horses, Yiosie2356, BG19bot, Dustin V. S., Ysjbserver, Mclaughtech and Anonymous: 2

- **Inspiration Mars Foundation** *Source:* https://en.wikipedia.org/wiki/Inspiration_Mars_Foundation?oldid=684569296 *Contributors:* Nealmcb, Patrick, Dhart, Bearcat, Axeman, Angry candy, Rich Farmbrough, Eric Kvaalen, Scriberius, BD2412, Drbogdan, Arado, Cmglee, Nickst, Skizzik, Chris the speller, Cattus, Derek R Bullamore, JorisvS, WilliamJE, N2e, Cydebot, VAXHeadroom, Andyjsmith, Ericmachmer, Callycrane, Z22, Magioladitis, KConWiki, BatteryIncluded, A3nm, CommonsDelinker, RockMFR, Wild8oar, Martarius, FOARP, Schreiber-Bike, Yobot, Againme, AnomieBOT, Invent2HelpAll, Fotaun, Brian the Editor, Shearonink, JoeSperrazza, Ypnypn, Noh Chung, BG19bot, Hz.tiang, ChrisGualtieri, Cerabot~enwiki, Reatlas, Ricks333, Randykitty, F6Zman, Wikiuser13, Suelru, Thiel217, Skyhook1, AndrewAntonio and Anonymous: 16

- **List of artificial objects on Venus** *Source:* https://en.wikipedia.org/wiki/List_of_artificial_objects_on_Venus?oldid=677840845 *Contributors:* Bryan Derksen, Wookie~enwiki, Reubenbarton, Cmapm, CharlesC, Drbogdan, Snappy, Nolanus, Poulpy, SmackBot, Nickst, Jeffro77, Papa November, Chlewbot, John, Novangelis, Checkguy, FairuseBot, Morganfitzp, Necessary Evil, Cydebot, Thijs!bot, Escarbot, Ricnun, GregU, Afterthewar~enwiki, CommonsDelinker, Rubble pile, RSStockdale, Gits (Neo), MystBot, Addbot, LinkFA-Bot, Luckas-bot, Xqbot, Fotaun, WikitanvirBot, ZéroBot, AvicAWB, Dexbot, Filedelinkerbot and Anonymous: 10

- **List of missions to Venus** *Source:* https://en.wikipedia.org/wiki/List_of_missions_to_Venus?oldid=694306474 *Contributors:* Smalljim, Smurrayinchester, WDGraham, Maddoug, Nikthestunned, Fotaun, Tom.Reding, EmausBot, ClueBot NG, Ninney, BattyBot, Dexbot, ToonLucas22, Günther gras88 and Anonymous: 3

- **Magellan (spacecraft)** *Source:* https://en.wikipedia.org/wiki/Magellan_(spacecraft)?oldid=685638843 *Contributors:* Mav, Bryan Derksen, Edward, Patrick, Ixfd64, Egil, Ahoerstemeier, Darkwind, Aarchiba, HarryHenryGebel, Robbot, Hadal, Curps, Ddama, Finn-Zoltan, Alexf, Superborsuk, ScottyBoy900Q, Urhixidur, Willhsmit, Deglr6328, N328KF, Wk muriithi, Bender235, BrokenSegue, Cwolfsheep, Giraffedata,

Supersexyspacemonkey, Kitch, Japanese Searobin, Bricktop, Chochopk, Smartech~enwiki, Drbogdan, Rjwilmsi, Tim!, BlueMoonlet, Teemu Maki~enwiki, SchuminWeb, Joedeshon, Dugo, Bgwhite, Whosasking, YurikBot, Conscious, SpuriousQ, Rapomon, Tony1, Jwissick, JoanneB, GrinBot~enwiki, Prvc, Anton n, SmackBot, Davepape, Eskimbot, Bluebot, Keegan, Scwlong, WDGraham, Andy120290, Vina-iwbot~enwiki, RandomCritic, Xiaphias, Rwboa22, Tobyw87, Novangelis, Joseph Solis in Australia, CmdrObot, Ruslik0, Korandder, Yaris678, Badseed, Cydebot, Vanished user vjhsduheuiui4t5hjri, Headbomb, Bobblehead, Dawnseeker2000, Escarbot, Ricnun, Pwhitwor, Terraracer, Volcanoguy, IanOsgood, WolfmanSF, JamesBWatson, BatteryIncluded, Miarmyguy, MartinBot, Ebellii, Alexander Patrakov, CommonsDelinker, CASfan, J.delanoy, Hans Dunkelberg, MarceloB, Youngjim, NewEnglandYankee, Juliancolton, VolkovBot, ABF, Eve Hall, Rei-bot, Truthanado, Alleborgo Bot, Gerakibot, Aklassyguy, Oxymoron83, Tombomp, Murlough23, Randy Kryn, Martarius, ClueBot, Arakunem, Excirial, Alexbot, Versus22, Hotcrocodile, Addbot, Fgnievinski, Tide rolls, Lightbot, Zorrobot, Luckas-bot, Fraggle81, Archon 2488, Piano non troppo, Kingpin13, Flewis, Citation bot, Frankenpuppy, ArthurBot, LilHelpa, Xqbot, Nagualdesign, Torquemada082, Physdragon, Pinethicket, Tom.Reding, Wickedsweetcake, RobertMfromLI, EmausBot, Allana rose, Cogiati, ClueBot NG, Xession, O.Koslowski, Bibcode Bot, Cgruda, BG19bot, AvocatoBot, Ninney, Aisteco, Aditya.varshney08, BattyBot, Tutelary, Qrhoo, Nimesh Mistry, YFdyh-bot, EnzaiBot, Khazar2, Hmainsbot1, Saltlake, Aladdin Ali Baba, Jupiter-4, Jcpag2012, Praemonitus, Nreedy11, Monkbot, Tetra quark, Mwleeds and Anonymous: 109

- **Manned Venus Flyby** *Source:* https://en.wikipedia.org/wiki/Manned_Venus_Flyby?oldid=665291960 *Contributors:* Kuralyov, Benbest, Rjwilmsi, TheRingess, Bubba73, CPColin, Mtu, Nickst, LouScheffer, COMPFUNK2, John, Euchiasmus, RandomCritic, Rwboa22, Joseph Solis in Australia, ThreeBlindMice, Cydebot, Andyjsmith, Ricnun, Mark Grant, Jatkins, Afterthewar~enwiki, Jon Holly, Srengel, VQuakr, Brett Buck, Excirial, FOARP, Muro Bot, Graham1973, RightGot, Dthomsen8, 68Kustom, Addbot, Luckas-bot, AnomieBOT, Gigemag76, Nasa-verve, John of Reading, Alvez3, Pokbot, Ryan Vesey, Khazar2, EuroCarGT, Fox2k11 and Anonymous: 14

- **Mariner 1** *Source:* https://en.wikipedia.org/wiki/Mariner_1?oldid=683990722 *Contributors:* The Epopt, XJaM, Nealmcb, Ixfd64, Big Bob the Finder, Furrykef, AlexPlank, Alexwcovington, Curps, JamesHoadley, Rich Farmbrough, El C, Svdmolen, BrokenSegue, Malcolm rowe, Bill-Cook, Deacon of Pndapetzim, RJFJR, Japanese Searobin, Triddle, BlaiseFEgan, DePiep, Melesse, Rjwilmsi, Tim!, Nightscream, Salix alba, Bubba73, FlaBot, Who, Ewlyahoocom, Chobot, Van der Hoorn, Hihihi, Emilio floris, GrinBot~enwiki, 8472, AndersL, SmackBot, Mmernex, KnowledgeOfSelf, Albanynewyork1, KocjoBot~enwiki, Bluebot, McNeight, WDGraham, Fuhghettaboutit, John, Novangelis, MTSbot~enwiki, Craigboy, Jaksmata, Tawkerbot2, Rdunn, N2e, Cydebot, Thijs!bot, Barticus88, Wikid77, Bobblehead, Michaelocc, Yakushima, Mike Payne, Duckysmokton, Ebellii, R'n'B, Bunker by, J.delanoy, VolkovBot, JhsBot, Gerakibot, Murlough23, Mr. Granger, ClueBot, Alexbot, Gerhardvalentin, NellieBly, MystBot, Addbot, Mabdul, Ronhjones, Ka Faraq Gatri, Lightbot, Luckas-bot, Yobot, Fraggle81, KamikazeBot, AnomieBOT, RadioBroadcast, Lee6597, Dogposter, DrilBot, EmausBot, CaptRik, Stephen C Wells, SporkBot, CocuBot, Xession, Frze, Nimesh Mistry, EagerToddler39, Décima-decepción, SteenthIWbot, Jupiter-4, JaconaFrere and Anonymous: 38

- **Mariner 10** *Source:* https://en.wikipedia.org/wiki/Mariner_10?oldid=688468409 *Contributors:* The Epopt, Bryan Derksen, XJaM, William Avery, SimonP, Patrick, Zanimum, Alfio, Ghewgill, Pizza Puzzle, Chrisjj, Reubenbarton, Curps, JamesHoadley, Maver1ck, Gadfium, The Singing Badger, Rdsmith4, Urhixidur, Crystal Matrix, O'Dea, Rich Farmbrough, Stampedem, Charm, Remember, Svdmolen, BrokenSegue, Shenme, BillCook, A2Kafir, Stephen G. Brown, Tablizer, Dirac1933, K3rb, Kazvorpal, EasyTarget, Carcharoth, Eyreland, Fxer, Driftwoodzebulin, Tim!, Mike s, Lendorien, Sferrier, Ttwaring, Cjpuffin, FlaBot, Margosbot~enwiki, Chobot, ScottAlanHill, YurikBot, Epolk, Shawn81, Mythsearcher, Gaius Cornelius, Los688, Masamunecyrus, Gadget850, Robost, OunKaiHiong, GrinBot~enwiki, Sardanaphalus, SmackBot, YellowMonkey, Bazza 7, Jrockley, Chris the speller, MalafayaBot, Colonies Chris, Redline, WDGraham, Can't sleep, clown will eat me, Andy120290, Ohconfucius, SashatoBot, CFLeon, Erimus, Iliev, JorisvS, RandomCritic, Xiaphias, Keycard, Joseph Solis in Australia, Tawkerbot2, ThreeBlindMice, Ruslik0, N2e, Cydebot, Caliga10, Arb, JamesAM, Thijs!bot, Epbr123, Headbomb, Bobblehead, Escarbot, AntiVandalBot, WinBot, Ricnun, Altamel, JAnDbot, Rothorpe, WolfmanSF, Pedro, LowEarthOrbit, Doesper, CommonsDelinker, Pdeitiker, J.delanoy, Trusilver, Rod57, Rwessel, Idioma-bot, ACSE, VolkovBot, EchoBravo, Brewbooks, Sdsds, TXiKiBoT, EuTuga, Hqb, TheresJamInTheHills, Ferengi, LeaveSleaves, Raryel, SlipperyHippo, SieBot, Scarian, Gerakibot, Spectre9, TrufflesTheLamb, Murlough23, JL-Bot, Martarius, MBK004, ClueBot, Jerry Wright, TarzanASG, Eeekster, Arjayay, LobStoR, XLinkBot, Yyhglouy666, Harlock81, Kbdankbot, Addbot, Freakmighty, Goatmanxero, Doniago, Lightbot, Zorrobot, Luckas-bot, Amirobot, Rks13, Stearnesy33, Xqbot, Capricorn42, Nasnema, Hellstricken, Astro Reeves, Fotaun, FrescoBot, Anassagora, Phoenix7777, IVAN3MAN, Double sharp, Comet Tuttle, John of Reading, Rami radwan, Gwillhickers, Bastian964, ZéroBot, Special jew, Wieralee, Sven Manguard, ClueBot NG, Xession, BG19bot, Ninney, Dexbot, Artyfinkle, Joeinwiki, Jupiter-4, Jcpag2012, URMOM123, Balon Greyjoy, Thibaut120094, Isambard Kingdom, HippocampusPotamus and Anonymous: 91

- **Mariner 2** *Source:* https://en.wikipedia.org/wiki/Mariner_2?oldid=672650234 *Contributors:* The Epopt, Bryan Derksen, XJaM, Infrogmation, Jeffq, AlexPlank, DavidFisher, Alexwcovington, Reubenbarton, Curps, JamesHoadley, Ddama, ChicXulub, The Singing Badger, ALE!, Sam Hocevar, Urhixidur, DMG413, Rich Farmbrough, Kaszeta, Svdmolen, BrokenSegue, BillCook, Tablizer, AnnaP, Cmapm, Dave.Dunford, Adrian.benko, DonPMitchell, Carcharoth, Dysepsion, Lasunncty, Rjwilmsi, Tim!, MZMcBride, Lendorien, Bubba73, FlaBot, Margosbot~enwiki, Kolbasz, D.brodale, Guanxi, Chobot, Arado, Van der Hoorn, NawlinWiki, Howcheng, GrinBot~enwiki, Anton n, AndersL, SmackBot, YellowMonkey, KnowledgeOfSelf, Albanynewyork1, Doc Strange, Rearden Metal, Mego2005, Bruce Marlin, WDGraham, TheGerm, Andy120290, Wen D House, Vgy7ujm, RandomCritic, Xiaphias, RekishiEJ, JerryStone, Rdunn, ThreeBlindMice, N2e, Cydebot, Gogo Dodo, Missvain, Klausness, Rees11, WinBot, Ricnun, Obeattie, JAnDbot, .anacondabot, Yakushima, Swpb, K95, Cyktsui, Spellmaster, Ebellii, R'n'B, Pdeitiker, J.delanoy, Sdsds, Garyms1963, Gerakibot, Mickea, Murlough23, ClueBot, EoGuy, MystBot, Kbdankbot, Addbot, DOI bot, Ka Faraq Gatri, LaaknorBot, Luckas-bot, Legobot II, Amirobot, Spacebuff, AnomieBOT, RadioBroadcast, ArthurBot, Capricorn42, Lee6597, Jc3s5h, Rafael323, Tom.Reding, Full-date unlinking bot, Robvanvee, عباد مجاهد ديراني، Hajatvrc, WikitanvirBot, JustinTime55, ChiZeroOne, Defan24, DJDunsie, El Roih, CocuBot, O.Koslowski, Helpful Pixie Bot, Bibcode Bot, Kelseymh, Tarcil, Fylbecatulous, BattyBot, Tech77, Lugia2453, JediCouncilMemberScooter, Jupiter-4, Jcpag2012, InAndOutLand, Monkbot and Anonymous: 38

- **Mariner 5** *Source:* https://en.wikipedia.org/wiki/Mariner_5?oldid=680776112 *Contributors:* The Epopt, Mav, Bryan Derksen, Frecklefoot, Tim Starling, Romanm, Curps, JamesHoadley, Frencheigh, CryptoDerk, The Singing Badger, Urhixidur, Rich Farmbrough, Clawed, BrokenSegue, BillCook, DonPMitchell, Rjwilmsi, Tim!, Lendorien, FlaBot, Margosbot~enwiki, Chobot, YurikBot, Van der Hoorn, Los688, GrinBot~enwiki, SmackBot, Gilliam, WDGraham, Andy120290, RandomCritic, Xiaphias, Rwboa22, MTSbot~enwiki, Negadrive~enwiki, ThreeBlindMice, N2e, Cydebot, Tawkerbot4, Thijs!bot, AntiVandalBot, WinBot, Ricnun, Altamel, Storkk, JAnDbot, Magioladitis, Adrian J. Hunter, Pdeitiker, Warut, TXiKiBoT, SieBot, Gerakibot, Murlough23, Rlbarton, Graham1973, Kbdankbot, Addbot, Lightbot, Luckas-bot, GrouchoBot, Lee6597, Mnmngb, D'ohBot, Rafael323, WikitanvirBot, El Roih, CocuBot, CBHikes and Anonymous: 16

- **Mariner program** *Source:* https://en.wikipedia.org/wiki/Mariner_program?oldid=684612999 *Contributors:* RjLesch, The Epopt, Bryan Derksen, Andre Engels, XJaM, Tim Starling, Paul A, Alfio, Muriel Gottrop~enwiki, Andres, Morwen, Nickshanks, Robbot, Baldhur, JamesHoadley,

Cam, The Singing Badger, ALE!, Maximaximax, Urhixidur, Jkeiser, JTN, Jkl, Rich Farmbrough, Aranel, Kwamikagami, Pilatus, Remember, Svdmolen, Leif, KJK::Hyperion, Simone, Bricktop, Graham87, MZMcBride, Krash, Margosbot~enwiki, Chobot, NTBot~enwiki, RussBot, Hede2000, Van der Hoorn, טרול רפאים, Tony1, Tryphiodorus, Uach, Argo Navis, GrinBot~enwiki, GunnerJr, Sardanaphalus, AndersL, Unyoyega, Eskimbot, Scwlong, WDGraham, Andy120290, Ericl, Yatpay, Iliev, NewTestLeper79, Nobunaga24, MTSbot~enwiki, CmdrObot, ThreeBlindMice, N2e, Friendlystar, Cydebot, Kanags, Ledzeppelin321295, Dawnseeker2000, Ricnun, Storkk, JAnDbot, Cyktsui, Hdt83, PatGund, VolkovBot, Kentholke, Sdsds, Baileypalblue, BotKung, Billinghurst, AlleborgoBot, SieBot, Gerakibot, Nipsonanomhmata, Randy Kryn, MBK004, Unbuttered Parsnip, Coinmanj, BodhisattvaBot, Addbot, Roentgenium111, Fgnievinski, Navy blue84, AnomieBOT, RadioBroadcast, ArthurBot, Obersachsebot, Xqbot, Jimmarsmars, Fotaun, FrescoBot, OspreyPL, Wstrwald, Craig Pemberton, OgreBot, Tom.Reding, RedBot, عباد مجاهد ديراني, EmausBot, WikitanvirBot, RawrMage, Evanh2008, ZéroBot, AvicAWB, ClueBot NG, Widr, Helpful Pixie Bot, Jsbarry15, ChrisGualtieri, Jcpag2012, Anythingcouldhappen and Anonymous: 56

- **Pioneer Venus project** *Source:* https://en.wikipedia.org/wiki/Pioneer_Venus_project?oldid=689912318 *Contributors:* Bryan Derksen, Timo Honkasalo, Ed Poor, SimonP, Msablic, Stone, Audin, JonathanDP81, AnonMoos, Twang, Pigsonthewing, Ancheta Wis, Fleminra, Curps, JamesHoadley, RetiredUser2, Bumm13, Ary29, Tsemii, Rich Farmbrough, Mbini, BrokenSegue, Tablizer, Arthena, Kitch, Oleg Alexandrov, Bricktop, Drbogdan, Tim!, Krash, FlaBot, Mirecki, Gaius Cornelius, GrinBot~enwiki, Anton n, SmackBot, Winterheart, WDGraham, Andy120290, Nakon, ArglebargleIV, John, Vgy7ujm, RandomCritic, Xiaphias, Novangelis, Joseph Solis in Australia, Necessary Evil, Cydebot, Bobblehead, Ricnun, JAnDbot, Poolio, Afterthewar~enwiki, Martial75, VolkovBot, TXiKiBoT, Misterzh, Humane Earth, Murlough23, Randy Kryn, The Thing That Should Not Be, Pline, Addbot, Lightbot, Beechs, Legobot, Luckas-bot, Yobot, TaBOT-zerem, BentleyCoon, Lkt1126, Xqbot, Fotaun, Thehelpfulbot, Rafael323, Dinamik-bot, Alph Bot, EmausBot, Rami radwan, A2soup, Rickray777, Kholodovsky, El Roih, Cgruda, BendelacBOT, Ninney, 4throck, 4Jays1034, Armanjafari, Hmainsbot1, Cerabot~enwiki, Quickscan1, Farrajak and Anonymous: 22

- **Pioneer Venus Multiprobe** *Source:* https://en.wikipedia.org/wiki/Pioneer_Venus_Multiprobe?oldid=687422283 *Contributors:* WDGraham, PointyOintment, N2e, Randy Kryn, Fotaun, El Roih, Pietro13, HippocampusPotamus and Anonymous: 2

- **Pioneer Venus Orbiter** *Source:* https://en.wikipedia.org/wiki/Pioneer_Venus_Orbiter?oldid=689552661 *Contributors:* Bryan Derksen, Wavelength, WDGraham, PointyOintment, N2e, Cydebot, Randy Kryn, Fgnievinski, AnomieBOT, Fotaun, Tom.Reding, Wikipelli, A2soup, El Roih, Ninney, Khazar2, Jupiter-4, DavidLeighEllis, HippocampusPotamus and Anonymous: 5

- **TMK** *Source:* https://en.wikipedia.org/wiki/TMK?oldid=549466084 *Contributors:* Altenmann, Geni, RJHall, DonPMitchell, Rjwilmsi, Mike s, Bubba73, Ttwaring, Gurch, Zwobot, Eclipsed, TMK94, Ultramandk, Nickst, Bluebot, TheGerm, Andy120290, DMacks, RandomCritic, Novangelis, Peyre, Joseph Solis in Australia, Necessary Evil, Cydebot, N5iln, Ricnun, VoABot II, Afterthewar~enwiki, Nono64, Trashbag, The Mighty Kinkle, JukoFF, SVO, Afernand74, MystBot, Addbot, Tide rolls, Lightbot, Klingon83, Fotaun, Luis1414, 12cabrera, EmausBot and Anonymous: 7

- **Tyazhely Sputnik** *Source:* https://en.wikipedia.org/wiki/Tyazhely_Sputnik?oldid=664514660 *Contributors:* Bryan Derksen, Andre Engels, SimonP, Imran, Dimadick, Murzun, Urhixidur, Rich Farmbrough, Cwolfsheep, Hooperbloob, Li-sung, Vegaswikian, FlaBot, Fram, Geoffrey.landis, SmackBot, Hibernian, WDGraham, Proofreader, Andy120290, Addshore, RandomCritic, Novangelis, Chmee2, Cydebot, Tec15, Creamaster, CTZMSC3, Ricnun, JAnDbot, Afterthewar~enwiki, J.delanoy, Tom Paine, Kyle the bot, TXiKiBoT, Ktr101, Addbot, LaaknorBot, Numbo3-bot, Punux, Legobot, Luckas-bot, Yobot, JackieBot, GrouchoBot, Erik9bot, D'ohBot, Rafael323, Tom.Reding, Lotje, EmausBot, Somebody500, ZéroBot, ChuispastonBot, Tony Mach and Anonymous: 7

- **Vega 1** *Source:* https://en.wikipedia.org/wiki/Vega_1?oldid=681757504 *Contributors:* Bryan Derksen, Robbot, Kuralyov, Rich Farmbrough, Moanzhu, Adrian.benko, Bkkbrad, CharlesC, BD2412, Drbogdan, Tim!, Toxygen, Howcheng, SmackBot, Nickst, WDGraham, Andy120290, Shaolin128, CFLeon, RandomCritic, Xiaphias, ThreeBlindMice, Ruslik0, N2e, Cydebot, Tec15, Dtgriscom, Ricnun, JAnDbot, Magioladitis, MartinBot, CommonsDelinker, Ling8zhi, Murlough23, RSStockdale, Dthomsen8, Kbdankbot, Addbot, Captain-tucker, Luckas-bot, Fotaun, Full-date unlinking bot, Cnwilliams, Kaiomai, ZéroBot, Yowanvista, 4throck, MeanMotherJr, BattyBot, Faizan and Anonymous: 17

- **Vega 2** *Source:* https://en.wikipedia.org/wiki/Vega_2?oldid=690362335 *Contributors:* Bryan Derksen, Mdebets, Robbot, Kuralyov, Jkl, Rich Farmbrough, Bkkbrad, CharlesC, Driftwoodzebulin, BD2412, Drbogdan, Tim!, Toxygen, Gaius Cornelius, Nickst, Gilliam, WDGraham, Andy120290, CFLeon, RandomCritic, Xiaphias, Joseph Solis in Australia, ThreeBlindMice, Ruslik0, N2e, Cydebot, Tec15, Arb, Headbomb, Dtgriscom, Bobblehead, Escarbot, Ricnun, PhilKnight, Esoominim, CommonsDelinker, LordAnubisBOT, Potatoswatter, Murlough23, WikHead, Kbdankbot, Addbot, Ka Faraq Gatri, Yobot, Fotaun, Thehelpfulbot, Full-date unlinking bot, Jamietw, Kaiomai, ZéroBot, ClueBot NG, CocuBot, Yowanvista, 4throck, MeanMotherJr, Nimesh Mistry, YFdyh-bot, Marco.bs, Tannerismydog and Anonymous: 6

- **Venera** *Source:* https://en.wikipedia.org/wiki/Venera?oldid=689912450 *Contributors:* AxelBoldt, Bryan Derksen, Jeronimo, Rmhermen, Seldon~enwiki, Gbleem, Minesweeper, Ahoerstemeier, Notheruser, TUF-KAT, Jniemenmaa, Hawthorn, Sbwoodside, Tempshill, AlexPlank, Robbot, Chris Roy, Postdlf, Kneiphof, Hadal, Epton, Curps, Zaha, Petrol~enwiki, RetiredUser2, Bumm13, CALR, Rich Farmbrough, Zscout370, Worldtraveller, Cmdrjameson, .:Ajvol:., Polylerus, Avian, Santiparam, Rwendland, Wtmitchell, Grenavitar, Cmapm, Kitch, Adrian.benko, DonPMitchell, Bricktop, Smartech~enwiki, Emerson7, RuM, Chupon, Qwertyus, Drbogdan, Rjwilmsi, Koavf, MZMcBride, ScottJ, Bubba73, JanSuchy, FlaBot, SchuminWeb, MacRusgail, Mirecki, EamonnPKeane, YurikBot, TexasAndroid, NTBot~enwiki, Arado, Hede2000, Gaius Cornelius, Aeusoes1, Mmenal, Howcheng, DeadEyeArrow, Geoffrey.landis, GrinBot~enwiki, SmackBot, Mark Tranchant, Incnis Mrsi, Pgk, Chris the speller, Papa November, WDGraham, Neo139, LouScheffer, Andy120290, JorisvS, RandomCritic, A. Parrot, Rwboa22, Novangelis, Peyre, Elendal, Cydebot, Kanags, Tec15, ST47, PKT, Frozenport, Escarbot, Ricnun, Joe Schmedley, Ironiridis, Blaine Steinert, WolfmanSF, JNW, Soulbot, Afterthewar~enwiki, FisherQueen, CommonsDelinker, Francis Tyers, Tom Paine, Reedy Bot, Rod57, Phild 99, Zamonin, Offshell, Amikake3, Quentonamos, Hqb, Jen817, AlleborgoBot, SieBot, Ferret, Murlough23, MBK004, ChandlerMapBot, Hotcrocodile, Avoided, SilvonenBot, Addbot, AndersBot, Lightbot, Ivancurtisivancurtis, Greyhood, Luckas-bot, Electronsoup, RadioBroadcast, Citation bot, ArthurBot, GrouchoBot, RibotBOT, SassoBot, Tom.Reding, RedBot, ZéroBot, Wingman4l7, ChiZeroOne, ClueBot NG, Cwmhiraeth, Gareth Griffith-Jones, Bibcode Bot, Michael Barera, RichardMills65, YFdyh-bot, Selenagirl, Lugia2453, CYl7EPTEMA777, My name is not dave, Liz, Ornithoptera, Filedelinkerbot, Eudxa, Raizada Vivekanand, FlorZi and Anonymous: 95

- **Venera 1** *Source:* https://en.wikipedia.org/wiki/Venera_1?oldid=692981012 *Contributors:* Bryan Derksen, Andre Engels, Rmhermen, Imran, Ahoerstemeier, Hadal, Millosh, Curps, PFHLai, EugeneZelenko, Rich Farmbrough, .:Ajvol:., Gdavidp, Cmapm, DonPMitchell, Arnomane, Bricktop, LinkTiger, Smartech~enwiki, Tim!, Mike s, Mike Peel, SchuminWeb, YurikBot, Howcheng, Evrik, Geoffrey.landis, Curpsbot-unicodify, Chris93, SmackBot, Unyoyega, Eskimbot, WDGraham, LouScheffer, PhiJ, SashatoBot, JorisvS, RandomCritic, Xiaphias, Rdunn,

ThreeBlindMice, N2e, Cydebot, Tec15, Rgrof, Ebyabe, Ricnun, JAnDbot, ZackTheJack, Duckysmokton, Klackalica, CommonsDelinker, STBotD, VolkovBot, Sdsds, TXiKiBoT, Spinningspark, Arakunem, Mild Bill Hiccup, BOTarate, SilvonenBot, Kbdankbot, Addbot, Luckas-bot, Yobot, AnomieBOT, Rubinbot, Materialscientist, ArthurBot, GrouchoBot, Maasje, IVAN3MAN, EmausBot, John of Reading, Rami radwan, CocuBot, Widr, Nimesh Mistry, Mogism, CYl7EPTEMA777, TeJota24 and Anonymous: 17

- **Venera 7** *Source:* https://en.wikipedia.org/wiki/Venera_7?oldid=690663832 *Contributors:* Bryan Derksen, Rlandmann, Twang, Pigsonthewing, Jyril, Geni, Avihu, .:Ajvol:., Sam Korn, Mlm42, Gdavidp, Cmapm, DonPMitchell, Bricktop, Smartech~enwiki, RuM, Rjwilmsi, Tim!, Bubba73, SchuminWeb, MacRusgail, Chobot, YurikBot, GLaDOS, Curpsbot-unicodify, GrinBot~enwiki, Mejor Los Indios, SmackBot, WDGraham, Trekphiler, Andy120290, Aleator, RandomCritic, Xiaphias, Novangelis, Cydebot, Thijs!bot, Ricnun, Ebellii, CommonsDelinker, J.delanoy, Amikake3, Kyle the bot, Gruverja, Rubble pile, Suwatest, Faithlessthewonderboy, Church, Piledhigheranddeeper, Stepheng3, Aitias, Silvonen-Bot, Addbot, Download, Luckas-bot, Banjohunter, TParis, ArthurBot, Xqbot, Rafael323, Full-date unlinking bot, Jirka.h23, Rami radwan, ZéroBot, Ocaasi, Sonicyouth86, ClueBot NG, Aoke1989, El Roih, CocuBot, Ninney, LOT762, Joshtaco, CYI7EPTEMA777, Rfassbind, Anythingcouldhappen, HippocampusPotamus and Anonymous: 43

- **Venera 8** *Source:* https://en.wikipedia.org/wiki/Venera_8?oldid=683323142 *Contributors:* Bryan Derksen, Bobblewik, EugeneZelenko, Gdavidp, Cmapm, Gene Nygaard, DonPMitchell, Bricktop, Smartech~enwiki, Rjwilmsi, Tim!, FlaBot, SchuminWeb, Curpsbot-unicodify, SmackBot, WDGraham, Trekphiler, Tsca.bot, John, RandomCritic, Xiaphias, Novangelis, CaptainZond, Cydebot, Tec15, Thijs!bot, Headbomb, Ricnun, Magioladitis, Ebellii, VolkovBot, TXiKiBoT, Mangostar, Rubble pile, Le Pied-bot~enwiki, BOTarate, Addbot, Luckas-bot, ZK001, Rafael323, WikitanvirBot, Rami radwan, Aoke1989, El Roih, CocuBot, CYI7EPTEMA777, HippocampusPotamus and Anonymous: 8

- **Venera 9** *Source:* https://en.wikipedia.org/wiki/Venera_9?oldid=694470817 *Contributors:* Bryan Derksen, Rmhermen, Lee M, Twang, Rorro, Curps, Cam, Sam Hocevar, Heck~enwiki, Urhixidur, Storm Rider, Gdavidp, Cmapm, Gene Nygaard, DonPMitchell, Bricktop, Smartech~enwiki, Rjwilmsi, Tim!, Mike s, Splarka, SchuminWeb, Kummi, Van der Hoorn, Coderzombie, JLaTondre, Curpsbot-unicodify, KnightRider~enwiki, SmackBot, Kinhull, WDGraham, Andy120290, RandomCritic, Xiaphias, Novangelis, CaptainZond, Cydebot, Tec15, Ebyabe, Thijs!bot, Headbomb, AntiVandalBot, Ricnun, Magioladitis, Rod57, TXiKiBoT, Eskovan, ImageRemovalBot, Alexbot, BOTarate, Addbot, Luckas-bot, TaBOT-zerem, ZK001, AnomieBOT, RadioBroadcast, Xqbot, FrescoBot, Rafael323, Tom.Reding, Full-date unlinking bot, Comet Tuttle, Galactic Penguin SST, WikitanvirBot, Rami radwan, ZéroBot, ClueBot NG, El Roih, CocuBot, Ninney, Joeinwiki, DN-boards1, HippocampusPotamus, Daiza222 and Anonymous: 24

- **Venera 10** *Source:* https://en.wikipedia.org/wiki/Venera_10?oldid=683325514 *Contributors:* Bryan Derksen, Rmhermen, Lee M, Curps, Heck~enwiki, Tablizer, Gdavidp, Cmapm, DonPMitchell, Bricktop, Smartech~enwiki, Rjwilmsi, Tim!, Mike s, FlaBot, Splarka, SchuminWeb, YurikBot, Gaius Cornelius, Curpsbot-unicodify, SmackBot, RandomCritic, Xiaphias, Novangelis, CaptainZond, Schauf, Cydebot, Tec15, Thijs!bot, Headbomb, Ricnun, JAnDbot, Andonic, Magioladitis, TXiKiBoT, Rubble pile, ImageRemovalBot, DumZiBoT, Dark Mage, Silvonen-Bot, Addbot, Luckas-bot, Neurolysis, Thehelpfulbot, Io Herodotus, Rafael323, RedBot, Galactic Penguin SST, EmausBot, Primefac, ZéroBot, Ebrambot, El Roih, CocuBot, Ninney, CYI7EPTEMA777, HippocampusPotamus and Anonymous: 7

- **Venera 11** *Source:* https://en.wikipedia.org/wiki/Venera_11?oldid=683326648 *Contributors:* Bryan Derksen, Rmhermen, Jason Potter, Robbot, RedWolf, Curps, Bobblewik, Haham hanuka, AnnaP, Gdavidp, Cmapm, Bricktop, Kralizec!, Smartech~enwiki, Rjwilmsi, Tim!, Mike s, SchuminWeb, Curpsbot-unicodify, SmackBot, Delldot, WDGraham, Andy120290, JorisvS, RandomCritic, Xiaphias, Novangelis, CaptainZond, Cydebot, Tec15, Ricnun, Magioladitis, Doesper, Rod57, VolkovBot, TXiKiBoT, Rubble pile, Poop4Brains, BOTarate, Chaosdruid, Addbot, Luckas-bot, ZK001, Felipe P, Rubinbot, GrouchoBot, Tom.Reding, Full-date unlinking bot, Galactic Penguin SST, EmausBot, WikitanvirBot, Rami radwan, ZéroBot, El Roih, CYI7EPTEMA777, Filedelinkerbot, DN-boards1, HippocampusPotamus and Anonymous: 7

- **Venera 12** *Source:* https://en.wikipedia.org/wiki/Venera_12?oldid=683326717 *Contributors:* Bryan Derksen, Rmhermen, RedWolf, Bobblewik, Humblefool, Rich Farmbrough, Cmapm, Kralizec!, Smartech~enwiki, Rjwilmsi, Tim!, Mike s, SchuminWeb, Bisqwit, Curpsbot-unicodify, SmackBot, WDGraham, JorisvS, RandomCritic, Xiaphias, Novangelis, CaptainZond, N2e, Cydebot, Tec15, Headbomb, Ricnun, JAnDbot, Magioladitis, Doesper, TXiKiBoT, Rubble pile, BOTarate, Addbot, AndersBot, Luckas-bot, ZK001, ArthurBot, RibotBOT, Rafael323, Vicenarian, Full-date unlinking bot, Dinamik-bot, TjBot, Galactic Penguin SST, WikitanvirBot, Rami radwan, ZéroBot, El Roih, CocuBot, CYI7EPTEMA777, Filedelinkerbot, HippocampusPotamus and Anonymous: 8

- **Venera 13** *Source:* https://en.wikipedia.org/wiki/Venera_13?oldid=693553027 *Contributors:* Bryan Derksen, Cyde, Silvonen, BlueMars, Gdavidp, DonPMitchell, Edgerunner76, Jleon, Rjwilmsi, Tim!, SchuminWeb, SmackBot, ILBobby, Gilliam, MalafayaBot, Papa November, Rcbutcher, WDGraham, Andy120290, JorisvS, RandomCritic, Xiaphias, Novangelis, CaptainZond, N2e, Geekfox, Cydebot, Gogo Dodo, Tec15, ST47, Thijs!bot, Keraunos, Headbomb, Ricnun, Rhinoracer, Altamel, JAnDbot, Magioladitis, Cadsuane Melaidhrin, BatteryIncluded, CommonsDelinker, Chrismathewsjr, Monique34, Mlarocque, MADe, N7j, Z-angel~enwiki, Falkonry, DumZiBoT, Sixtyninefourtyninefourtyfoureleven, Addbot, Luckas-bot, Crispmuncher, AnomieBOT, Io Herodotus, Rafael323, Tom.Reding, RedBot, MastiBot, Full-date unlinking bot, Galactic Penguin SST, EmausBot, Rami radwan, Мирослав Ћика, AvicBot, ZéroBot, Makecat, Wingman4l7, L Kensington, El Roih, CocuBot, The rakish fellow, Jimyami, Lugia2453, CYI7EPTEMA777, Filedelinkerbot and Anonymous: 35

- **Venera 14** *Source:* https://en.wikipedia.org/wiki/Venera_14?oldid=693553542 *Contributors:* Bryan Derksen, Feedmecereal, BlueMars, DonPMitchell, Rjwilmsi, Tim!, SchuminWeb, 21655, SmackBot, Chris the speller, Rcbutcher, WDGraham, Andy120290, RandomCritic, Xiaphias, CaptainZond, Chrumps, N2e, Cydebot, Gogo Dodo, Tec15, Keraunos, Headbomb, Ricnun, JAnDbot, Magioladitis, Monique34, ClueBot, DumZiBoT, Sixtyninefourtyninefourtyfoureleven, Addbot, Luckas-bot, AnomieBOT, Felipe P, Almabot, GrouchoBot, Nicolas Perrault III, Tom.Reding, Full-date unlinking bot, Dinamik-bot, Galactic Penguin SST, Rami radwan, ZéroBot, ClueBot NG, Makecat-bot, CYI7EPTEMA777 and Anonymous: 20

- **Venera 15** *Source:* https://en.wikipedia.org/wiki/Venera_15?oldid=660801013 *Contributors:* Bryan Derksen, Silvonen, Gdavidp, Kitch, Rjwilmsi, Tim!, SmackBot, WDGraham, RandomCritic, Xiaphias, Novangelis, Cydebot, Tec15, Ricnun, JAnDbot, Magioladitis, Afterthewar~enwiki, Neonblak, TXiKiBoT, Z-angel~enwiki, MystBot, Addbot, Lightbot, Luckas-bot, ArthurBot, Almabot, GrouchoBot, HRoestBot, Tom.Reding, WikitanvirBot, El Roih, CocuBot, Oddbodz, CYI7EPTEMA777, Aladdin Ali Baba and Anonymous: 3

- **Venera 16** *Source:* https://en.wikipedia.org/wiki/Venera_16?oldid=691036572 *Contributors:* Bryan Derksen, Gdavidp, Rjwilmsi, Tim!, SmackBot, WDGraham, RandomCritic, Xiaphias, Novangelis, Cydebot, Tec15, Ricnun, JAnDbot, Magioladitis, Afterthewar~enwiki, Neonblak, Ebellii, Addbot, Lightbot, Luckas-bot, Quentinv57, AnomieBOT, Almabot, GrouchoBot, Tom.Reding, Dinamik-bot, ZéroBot, El Roih, CocuBot, YFdyh-bot, Tony Mach, CYI7EPTEMA777, Aladdin Ali Baba, HgandVenus, Pietro13 and Anonymous: 3

- **Venera 2MV-1 No.2** *Source:* https://en.wikipedia.org/wiki/Venera_2MV-1_No.2?oldid=586361349 *Contributors:* Mav, Bryan Derksen, Andre Engels, SimonP, Imran, Randomm~enwiki, Dimadick, FlaBot, Banes, Fram, SmackBot, WDGraham, Aces lead, RandomCritic, Novangelis, Cydebot, Ricnun, Afterthewar~enwiki, VolkovBot, Kresadlo, Andy Dingley, SieBot, Ktr101, DumZiBoT, Addbot, GDK, Luckas-bot, Yobot, Momoricks, Erik9bot, Rafael323, Beltline, Gisegre, ZéroBot, YFdyh-bot, JediCouncilMemberScooter and Anonymous: 1

- **Venera 2MV-2 No.1** *Source:* https://en.wikipedia.org/wiki/Venera_2MV-2_No.1?oldid=601714733 *Contributors:* Tim!, FlaBot, Tijuana Brass, Mholland, Deuar, SmackBot, WDGraham, Andy120290, RandomCritic, Novangelis, Cydebot, Tec15, Epbr123, Ricnun, VolkovBot, Kresadlo, SieBot, ClueBot, Ktr101, Addbot, LaaknorBot, GDK, Luckas-bot, Yobot, D'ohBot, Rafael323, Full-date unlinking bot, CocuBot and Anonymous: 4

- **Venera-D** *Source:* https://en.wikipedia.org/wiki/Venera-D?oldid=694176184 *Contributors:* Vegaswikian, FlaBot, EamonnPKeane, Aeusoes1, Geoffrey.landis, Sardanaphalus, Nickst, Cattus, MaxSem, Andy120290, JorisvS, RandomCritic, Rwboa22, Joseph Solis in Australia, Cydebot, Ricnun, BatteryIncluded, NERV~enwiki, Afterthewar~enwiki, Tinwelint, Tom Paine, Bill.jesdale, A4bot, Killing sparrows, AlleborgoBot, Kktor, RSStockdale, MBK004, DumZiBoT, Addbot, Lightbot, Vasiľ, Amirobot, Lord Aro, ArthurBot, DirlBot, Arsia Mons, Lankier, DrilBot, HRoestBot, Rami radwan, ZéroBot, Wingman4l7, CocuBot, BIT1982, Danim, BG19bot, Zetrs, BattyBot, Stilgar27 and Anonymous: 9

- **Venus Express** *Source:* https://en.wikipedia.org/wiki/Venus_Express?oldid=686756402 *Contributors:* Bryan Derksen, Danny, XJaM, Rmhermen, Mahjongg, Cyde, Mihai~enwiki, Stone, Jitse Niesen, Lumos3, Robbot, Astronautics~enwiki, RedWolf, PedroPVZ, Vasi, Awolf002, Cassioli, Harp, Maver1ck, Python eggs, Geni, The Singing Badger, FelineAvenger, Kaldari, Hellisp, Willhsmit, MartinBiely, Grstain, N328KF, JTN, A-giau, Rich Farmbrough, Kyknos, Deprifry, Huntster, Shanes, Iamunknown, Pocket Rockets, Hooperbloob, Hektor, Wricardoh~enwiki, Keenan Pepper, VladimirKorablin, Snowolf, Knowledge Seeker, Evil Monkey, Gene Nygaard, Dan100, Adrian.benko, Zntrip, MartinSpacek, Lost.goblin, Uncle G, BillC, Bricktop, RedBLACKandBURN, Li-sung, Rjwilmsi, Tim!, Bhadani, Marsbound2024, G Clark, Crazycomputers, Shadow007, Melancholie, Mirecki, Zotel, Narvalo, Chobot, Wjfox2005, YurikBot, Mithridates, Gaius Cornelius, Los688, Ravedave, Tony1, Dv82matt, Yonidebest, HDietze, Ageekgal, NHSavage, Hurricane Devon, Philip Stevens, That Guy, From That Show!, SG, SmackBot, TestPilot, Nickst, Jonathan Karlsson, Dogmatico (usurped), MalafayaBot, Utah~enwiki, WDGraham, Ww2censor, Derek R Bullamore, Jbergquist, Molerat, JorisvS, RandomCritic, Xiaphias, Novangelis, CmdrObot, ThreeBlindMice, N2e, Juhachi, Devatipan, Necessary Evil, Cydebot, Kanags, Citriczylaphone, Gogo Dodo, DMeyering, S Marshall, Bobblehead, KrakatoaKatie, Planetary, IanOsgood, Smihael, WODUP, BatteryIncluded, Winiar, Markco1, Nopira, MartinBot, R'n'B, CommonsDelinker, Pdeitiker, Hans Dunkelberg, Potatoswatter, Usp, Agamemnus, ACSE, 28bytes, Sdsds, SpikeZOM, JavierCastro, Fredibus, Murlough23, RSStockdale, Dillard421, MBK004, Anonymous799, Gits (Neo), Niceguyedc, Com4space, Panther303, PixelBot, Olivierclaudel, RubenGarciaHernandez, MelonBot, Addbot, DOI bot, LenardLenard, Ka Faraq Gatri, 37ophiuchi, Yobot, Eric-Wester, AnomieBOT, Bluerasberry, Citation bot, Xqbot, Fotaun, Cndwrld, 117Avenue, Citation bot 1, Diannaa, Pobcs, RjwilmsiBot, Bento00, EmausBot, John of Reading, Helium4, Dewritech, ZéroBot, Science enthusiast343, Kobbra, EvenGreenerFish, ChiZeroOne, ClueBot NG, Xession, Bennetscrose, Kohra, Widr, Ganman33, Hallows AG, PhilipTerryGraham, Mogism, Jc86035, Generic1139, Rfassbind, Space12345, Monkbot, Aarrcchhiimmeeddeess and Anonymous: 98

- **Venus In Situ Explorer** *Source:* https://en.wikipedia.org/wiki/Venus_In_Situ_Explorer?oldid=650986779 *Contributors:* Bender235, Alai, Lasunncty, ElfQrin, RadioFan, Tony1, Geoffrey.landis, SmackBot, Nickst, WDGraham, JorisvS, Joseph Solis in Australia, Ruslik0, Cydebot, Alaibot, BatteryIncluded, Million Moments, Shoemoney2night, T.Neo, Wwheaton, Groslard, DumZiBoT, Addbot, Takamachi5039, Luckas-bot, AnomieBOT, Materialscientist, ArthurBot, Xqbot, Fotaun, FrescoBot, EmausBot, WikitanvirBot, Rami radwan, GoingBatty, ZéroBot, CocuBot, Grimnir11, Nydoc1 and Anonymous: 9

- **Venus orbiter mission** *Source:* https://en.wikipedia.org/wiki/Venus_orbiter_mission?oldid=685567779 *Contributors:* Rpyle731, BatteryIncluded, Varmapak, BG19bot, Sulfurboy and Jayprakash12345

- **Venus Orbiting Imaging Radar** *Source:* https://en.wikipedia.org/wiki/Venus_Orbiting_Imaging_Radar?oldid=693087928 *Contributors:* RadioFan, Jodosma, Praemonitus and Ysjbserver

- **VERITAS (spacecraft)** *Source:* https://en.wikipedia.org/wiki/VERITAS_(spacecraft)?oldid=687975441 *Contributors:* JohnCD, BatteryIncluded, Fotaun and Anonymous: 1

- **Vesta (spacecraft)** *Source:* https://en.wikipedia.org/wiki/Vesta_(spacecraft)?oldid=634855175 *Contributors:* Drbogdan, Rjwilmsi, Mike Selinker, Nickst, WDGraham, Gildir, JorisvS, N2e, Cydebot, Tec15, Ricnun, Lightbot, AnomieBOT, Artvill, Шугуан, El Roih and Anonymous: 2

- **Zond 1** *Source:* https://en.wikipedia.org/wiki/Zond_1?oldid=664974549 *Contributors:* Bryan Derksen, Curps, Squash, Jkeiser, Qutezuce, DonPMitchell, Tim!, YurikBot, GrinBot~enwiki, SmackBot, WDGraham, Andy120290, Valenciano, RandomCritic, Xiaphias, CmdrObot, Cydebot, Tec15, Headbomb, Dtgriscom, Ricnun, Afterthewar~enwiki, CommonsDelinker, SieBot, RSStockdale, Z-angel~enwiki, BodhisattvaBot, Addbot, Алиса Селезнёва, Luckas-bot, Xqbot, Maasje, ZéroBot, Alvez3 and Anonymous: 4

- **3MV** *Source:* https://en.wikipedia.org/wiki/3MV?oldid=587540402 *Contributors:* Avian, Lockley, WDGraham, JHunterJ, Flowerpotman, Ohms law, Billinghurst, Malcolmxl5, Nn123645, ImageRemovalBot, Chaosdruid, Addbot, AnomieBOT, Xqbot, LucienBOT, Ripchip Bot, EmausBot, Wingman4l7, Alvez3 and Anonymous: 1

- **Vega program** *Source:* https://en.wikipedia.org/wiki/Vega_program?oldid=664246104 *Contributors:* Bryan Derksen, Qfwfq, Hike395, Robbot, Romanm, Ojigiri~enwiki, Curps, Bobblewik, RetiredUser2, Kuralyov, ArnoldReinhold, Huntster, Thuresson, Tjic, Avian, Grenavitar, Cmapm, Gene Nygaard, Kitch, Bricktop, Marudubshinki, Drbogdan, Daderot, Bovineone, Tony1, Curpsbot-unicodify, 8472, Anton n, Neier, Nickst, Chris the speller, Hibernian, Kactuswren, WDGraham, Chlewbot, Andy120290, Kuru, JorisvS, RandomCritic, Novangelis, CaptainZond, CmdrObot, Cydebot, Tec15, Thijs!bot, Headbomb, Dtgriscom, Ricnun, JAnDbot, CommonsDelinker, Coppertwig, Mstuomel, Redsunrising, DorganBot, Macboff, TXiKiBoT, SieBot, XLinkBot, Addbot, CanadianLinuxUser, Lightbot, Luckas-bot, Ptbotgourou, J04n, Surv1v4l1st, Tom.Reding, Full-date unlinking bot, Lotje, Dinamik-bot, RjwilmsiBot, EmausBot, Dewritech, Disambigutron, Шугуан, Wingman4l7, ChiZeroOne, CocuBot, Bibcode Bot, Cgruda, ElphiBot, Szczureq, SteenthIWbot, Pikador and Anonymous: 28

- **2V (V-69)** *Source:* https://en.wikipedia.org/wiki/2V_(V-69)?oldid=678141288 *Contributors:* BD2412, Anomalocaris, Malcolma, Cydebot, Pi.1415926535, AnomieBOT, Omnipaedista, AvicBot, Mrmatiko, El Roih and Anonymous: 1

- **Kosmos 21** *Source:* https://en.wikipedia.org/wiki/Kosmos_21?oldid=586338658 *Contributors:* AxelBoldt, Andre Engels, SimonP, Imran, Dimadick, Timrollpickering, Rich Farmbrough, SmackBot, Hmains, WDGraham, LouScheffer, RandomCritic, Novangelis, Bocianski, Cydebot, Tec15, Ricnun, Tom Paine, Phild 99, VolkovBot, TXiKiBoT, Alexbot, Addbot, Luckas-bot, Yobot, Rubinbot, Erik9bot, D'ohBot, Rafael323, ZéroBot, BattyBot and Anonymous: 1

- **Kosmos 27** *Source:* https://en.wikipedia.org/wiki/Kosmos_27?oldid=586339105 *Contributors:* SimonP, Imran, PhilipMW, Timrollpickering, Klemen Kocjancic, Xosé, WDGraham, RandomCritic, Novangelis, Cydebot, Ricnun, Belovedfreak, TXiKiBoT, SieBot, Addbot, Luckas-bot, Stellar Grifon, Rubinbot, GrouchoBot, Erik9bot, D'ohBot, MastiBot, ZéroBot and Anonymous: 1

- **Kosmos 167** *Source:* https://en.wikipedia.org/wiki/Kosmos_167?oldid=641349807 *Contributors:* Kralizec!, Qwertyus, WDGraham, N2e, Cydebot, AnomieBOT, Tom.Reding and Ninney

- **Kosmos 482** *Source:* https://en.wikipedia.org/wiki/Kosmos_482?oldid=634565257 *Contributors:* Timrollpickering, Grutness, Smoth 007, Rjwilmsi, Tim!, Whosasking, Nickst, Hmains, Bluebot, WDGraham, LouScheffer, RandomCritic, Cydebot, Tec15, Ricnun, Afterthewar~enwiki, Ebellii, TXiKiBoT, SieBot, Kbdankbot, Addbot, Tothwolf, Luckas-bot, Yobot, AnomieBOT, Full-date unlinking bot, EmausBot, ZéroBot, Jupiter-4 and Anonymous: 2

- **Kosmos 96** *Source:* https://en.wikipedia.org/wiki/Kosmos_96?oldid=629618648 *Contributors:* Jayjg, Kralizec!, WDGraham, Cydebot, JediCouncilMemberScooter and Anonymous: 1

- **Venera 2** *Source:* https://en.wikipedia.org/wiki/Venera_2?oldid=655859546 *Contributors:* Bryan Derksen, Cmapm, Bricktop, Smartech~enwiki, Rjwilmsi, Tim!, Rui Silva, FlaBot, SchuminWeb, Roboto de Ajvol, YurikBot, Mmenal, Curpsbot-unicodify, GrinBot~enwiki, SmackBot, Unyoyega, WDGraham, Rodri316, RandomCritic, Xiaphias, Novangelis, N2e, Cydebot, Tec15, Thijs!bot, Ricnun, VolkovBot, TXiKiBoT, BOTarate, Addbot, Luckas-bot, Yobot, Ptbotgourou, Xqbot, Erik9bot, Fotaun, Rami radwan, ZéroBot, CYl7EPTEMA777 and Anonymous: 6

- **Venera 3** *Source:* https://en.wikipedia.org/wiki/Venera_3?oldid=651659746 *Contributors:* Mav, Bryan Derksen, Rmhermen, Ahoerstemeier, Robbot, Curps, EugeneZelenko, .:Ajvol:., Helix84, TommyBoy, Gdavidp, Cmapm, Adrian.benko, Bricktop, Zzyzx11, Smartech~enwiki, Rjwilmsi, Tim!, Mike s, Rui Silva, SLi, FlaBot, SchuminWeb, Roboto de Ajvol, YurikBot, Mmenal, Curpsbot-unicodify, GrinBot~enwiki, SmackBot, Unyoyega, WDGraham, Proofreader, Andy120290, Nakon, SashatoBot, RandomCritic, Xiaphias, Novangelis, Dl2000, Cydebot, Tec15, Ricnun, Armking2, VolkovBot, A4bot, SieBot, Rubble pile, Excirial, Canis Lupus, Doprendek, Muro Bot, BOTarate, Addbot, Luckas-bot, Yobot, Materialscientist, Full-date unlinking bot, Rami radwan, ZéroBot, राग, Ninney, BOBTUB, Nimesh Mistry, CYl7EPTEMA777 and Anonymous: 13

- **Venera 4** *Source:* https://en.wikipedia.org/wiki/Venera_4?oldid=693638181 *Contributors:* Bryan Derksen, Twilsonb, Geni, TommyBoy, Cmapm, Adrian.benko, DonPMitchell, Bricktop, Smartech~enwiki, Qwertyus, Rjwilmsi, Tim!, Mike Peel, FlaBot, SchuminWeb, MacRusgail, Kummi, YurikBot, -OOPSIE-, Thalter, Curpsbot-unicodify, GrinBot~enwiki, SmackBot, WDGraham, Tsca.bot, LouScheffer, RandomCritic, Xiaphias, Novangelis, IronJohnSr, Cydebot, Hebrides, Tec15, Ricnun, WolfmanSF, Trusilver, Warut, VolkovBot, Kyle the bot, TXiKiBoT, Burntsauce, MikeGruz, Piledhigheranddeeper, BOTarate, Addbot, Favonian, Luckas-bot, TheSuave, Materialscientist, ArthurBot, Mnmngb, Jlg3926, Sjolden, AA5L, RjwilmsiBot, Polylepsis, ZéroBot, CocuBot, Helpful Pixie Bot, Gob Lofa, Bibcode Bot, CYl7EPTEMA777, Ruby Murray, Filedelinkerbot, Esquetially, Terklory7 and Anonymous: 28

- **Venera 5** *Source:* https://en.wikipedia.org/wiki/Venera_5?oldid=619968627 *Contributors:* Bryan Derksen, Slawojarek, PFHLai, Grm wnr, Slipstream, .:Ajvol:., Gdavidp, Cmapm, Adrian.benko, DonPMitchell, Bricktop, Smartech~enwiki, RuM, Rjwilmsi, Tim!, FlaBot, SchuminWeb, Roboto de Ajvol, YurikBot, Gaius Cornelius, Curpsbot-unicodify, SmackBot, WDGraham, Aces lead, John, RandomCritic, Xiaphias, Novangelis, Cydebot, Tec15, Headbomb, Ricnun, Magioladitis, TXiKiBoT, Rubble pile, BOTarate, SilvonenBot, Addbot, Mnh, Luckas-bot, X2ca, Xqbot, Io Herodotus, Inscription, Full-date unlinking bot, WikitanvirBot, Rami radwan, ZéroBot, CocuBot, Ninney, 220 of Borg, CYl7EPTEMA777 and Anonymous: 11

- **Venera 6** *Source:* https://en.wikipedia.org/wiki/Venera_6?oldid=607722055 *Contributors:* Bryan Derksen, PaulinSaudi, Thue, Curps, Ormondad, .:Ajvol:., Gdavidp, Cmapm, DonPMitchell, E. Brown, Bricktop, Smartech~enwiki, Rjwilmsi, Tim!, FlaBot, SchuminWeb, Roboto de Ajvol, YurikBot, Curpsbot-unicodify, SmackBot, WDGraham, John, RandomCritic, Xiaphias, Novangelis, Cydebot, Tec15, Headbomb, Ricnun, JAnDbot, Magioladitis, MartinBot, Rubble pile, BOTarate, Addbot, Crazy Ivan, Luckas-bot, ArthurBot, Xqbot, LucienBOT, Io Herodotus, D'ohBot, Rafael323, Full-date unlinking bot, Rami radwan, Aoke1989, CocuBot, Ninney, CYl7EPTEMA777 and Anonymous: 8

- **Venera 2MV-1 No.1** *Source:* https://en.wikipedia.org/wiki/Venera_2MV-1_No.1?oldid=586361337 *Contributors:* Andre Engels, Imran, Bevo, Robbot, Tualha, Rich Farmbrough, Tim!, Shauri, FlaBot, Deuar, SmackBot, WDGraham, RandomCritic, Novangelis, Cydebot, Tec15, Escarbot, Ricnun, Afterthewar~enwiki, Rocketmaniac, VolkovBot, Sdsds, TXiKiBoT, Kresadlo, SieBot, Ktr101, DumZiBoT, Addbot, GDK, Luckas-bot, Yobot, GrouchoBot, CocuBot, JediCouncilMemberScooter and Anonymous: 1

- **Venera 4V-2** *Source:* https://en.wikipedia.org/wiki/Venera_4V-2?oldid=660801177 *Contributors:* Mav, Bryan Derksen, Rmhermen, Heron, Ahoerstemeier, Tristanb, Naddy, Curps, Bobblewik, Grenavitar, Cmapm, DonPMitchell, Bricktop, Smartech~enwiki, Rjwilmsi, Mike s, Curpsbot-unicodify, SmackBot, WDGraham, Andy120290, RandomCritic, Novangelis, Cydebot, Ricnun, JAnDbot, Txomin, Magioladitis, Afterthewar~enwiki, Jorfer, JL-Bot, WikHead, Addbot, AndersBot, RibotBOT, Rafael323, Tom.Reding, Wingman4l7, Aladdin Ali Baba and Anonymous: 3

## 2.2  Images

- **File:1986_venera_galley_nh.jpg** *Source:* https://upload.wikimedia.org/wikipedia/commons/8/8d/1986_venera_galley_nh.jpg *License:* Public domain *Contributors:* http://russianstamps.ru/ *Original artist:* Post of Soviet Union

- **File:Addams_crater_on_Venus.jpg** *Source:* https://upload.wikimedia.org/wikipedia/commons/7/73/Addams_crater_on_Venus.jpg *License:* Public domain *Contributors:* http://www.espacial.org/images/jpg/addams.jpg *Original artist:* NASA's Magellan probe

- **File:Inspiration_Mars_Periapsis.jpg** *Source:* https://upload.wikimedia.org/wikipedia/commons/8/8f/Inspiration_Mars_Periapsis.jpg *License:* CC BY-SA 2.0 *Contributors:* http://www.flickr.com/photos/94053242@N04/8557315459/in/photostream *Original artist:* Inspiration Mars Foundation

- **File:Inspiration_Mars_trajectory.svg** *Source:* https://upload.wikimedia.org/wikipedia/commons/4/42/Inspiration_Mars_trajectory.svg *License:* CC BY-SA 3.0 *Contributors:* Own work *Original artist:* Cmglee

- **File:Isabella_Crater_PIA00480.jpg** *Source:* https://upload.wikimedia.org/wikipedia/commons/8/87/Isabella_Crater_PIA00480.jpg *License:* Public domain *Contributors:* http://photojournal.jpl.nasa.gov/catalog/PIA00480 *Original artist:* Magellan probe

- **File:KSC-67PC-0184.jpg** *Source:* https://upload.wikimedia.org/wikipedia/commons/4/41/KSC-67PC-0184.jpg *License:* Public domain *Contributors:* Kennedy Space Center Photo Archive (image link) *Original artist:* NASA

- **File:Kennedy_Receives_Mariner_2_Model.jpg** *Source:* https://upload.wikimedia.org/wikipedia/commons/1/13/Kennedy_Receives_Mariner_2_Model.jpg *License:* Public domain *Contributors:* Great Images in NASA *Original artist:* NASA

- **File:Largevenusprobe.gif** *Source:* https://upload.wikimedia.org/wikipedia/en/6/6a/Largevenusprobe.gif *License:* Fair use *Contributors:* Annual Big Soviet Encyclopaedia Journal

  *Original artist:*

  Unknown

- **File:Local-time.svg** *Source:* https://upload.wikimedia.org/wikipedia/commons/7/7f/Local-time.svg *License:* CC BY-SA 2.5 *Contributors:* self-made Image:Applications-internet.svg Image:Ambox outdated content.svg *Original artist:* Xander

- **File:MESSENGER_-_Venus_630_nm_stretch.jpg** *Source:* https://upload.wikimedia.org/wikipedia/commons/3/3b/MESSENGER_-_Venus_630_nm_stretch.jpg *License:* Public domain *Contributors:* http://messenger.jhuapl.edu/gallery/sciencePhotos/image.php?page=1&gallery_id=2&image_id=327 *Original artist:* NASA / JHU/APL

- **File:Maat_Mons_on_Venus.jpg** *Source:* https://upload.wikimedia.org/wikipedia/commons/1/16/Maat_Mons_on_Venus.jpg *License:* Public domain *Contributors:* ? *Original artist:* ?

- **File:Magellan_-_Magellan_Spacecraft_in_Preflight_Checkout_at_Kennedy_Space_Center.png** *Source:* https://upload.wikimedia.org/wikipedia/commons/b/b5/Magellan_-_Magellan_Spacecraft_in_Preflight_Checkout_at_Kennedy_Space_Center.png *License:* Public domain *Contributors: Magellan: Mission to Venus,* JPL, 1991 *Original artist:* NASA / JPL

- **File:Magellan_-_Venera_10_landing_site_mgn_c115n283_1.gif** *Source:* https://upload.wikimedia.org/wikipedia/commons/f/f7/Magellan_-_Venera_10_landing_site_mgn_c115n283_1.gif *License:* Public domain *Contributors:* http://nssdc.gsfc.nasa.gov/imgcat/html/object_page/mgn_c115n283_1.html *Original artist:* NASA

- **File:Magellan_-_antennas.png** *Source:* https://upload.wikimedia.org/wikipedia/commons/1/1b/Magellan_-_antennas.png *License:* Public domain *Contributors: Magellan: Mission to Venus* [1], pg. 8 *Original artist:* NASA / JPL

- **File:Magellan_-_artist_depiction.png** *Source:* https://upload.wikimedia.org/wikipedia/commons/6/68/Magellan_-_artist_depiction.png *License:* Public domain *Contributors:* http://www2.jpl.nasa.gov/magellan/images.html *Original artist:* NASA

- **File:Magellan_-_attitude_and_propulsion.png** *Source:* https://upload.wikimedia.org/wikipedia/commons/8/80/Magellan_-_attitude_and_propulsion.png *License:* Public domain *Contributors: Magellan: Mission to Venus* [1], pg. 10 / 12 *Original artist:* NASA / JPL

- **File:Magellan_-_burst_rate_diagram_-_orig.png** *Source:* https://upload.wikimedia.org/wikipedia/commons/c/c9/Magellan_-_burst_rate_diagram_-_orig.png *License:* Public domain *Contributors:* Dallas, S.S. *THE VENUS RADAR MAPPER MISSION,* JPL, 1987 *Original artist:* NASA / JPL

- **File:Magellan_-_cycle_1_map_-_1298972763463430062.580622.jpg** *Source:* https://upload.wikimedia.org/wikipedia/commons/7/79/Magellan_-_cycle_1_map_-_1298972763463430062.580622.jpg *License:* Public domain *Contributors:* http://www.mapaplanet.org/explorer/maps/1298972763463430062.580622.jpg *Original artist:* NASA / USGS

- **File:Magellan_-_cycle_2_map_-_1299005988110004007.109767.jpg** *Source:* https://upload.wikimedia.org/wikipedia/commons/a/a5/Magellan_-_cycle_2_map_-_1299005988110004007.109767.jpg *License:* Public domain *Contributors:* http://www.mapaplanet.org/explorer/maps/1299005988110004007.109767.jpg *Original artist:* NASA / USGS

- **File:Magellan_-_cycle_3_map_-_1299006020759454362.039599.jpg** *Source:* https://upload.wikimedia.org/wikipedia/commons/6/66/Magellan_-_cycle_3_map_-_1299006020759454362.039599.jpg *License:* Public domain *Contributors:* http://www.mapaplanet.org/explorer/maps/1299006020759454362.039599.jpg *Original artist:* NASA / USGS

- **File:Magellan_-_data_gathering_diagram.png** *Source:* https://upload.wikimedia.org/wikipedia/commons/f/f3/Magellan_-_data_gathering_diagram.png *License:* Public domain *Contributors:* Pettengill, *Magellan: Radar Performance and Data Products,* AAAS, JPL, 1991, p. 262 *Original artist:* NASA / JPL

- **File:Magellan_-_diagram_of_atimetry_and_SAR_data_gathering.png** *Source:* https://upload.wikimedia.org/wikipedia/commons/3/34/Magellan_-_diagram_of_atimetry_and_SAR_data_gathering.png *License:* Public domain *Contributors: Magellan: Mission to Venus,* JPL, 1991 *Original artist:* NASA / JPL

- **File:Magellan_-_end_of_mission_poster_-_mgnlogo2.gif** *Source:* https://upload.wikimedia.org/wikipedia/commons/8/87/Magellan_-_end_of_mission_poster_-_mgnlogo2.gif *License:* Public domain *Contributors:* http://www2.jpl.nasa.gov/magellan/status.html Direct link *Original artist:* NASA/JPL

- **File:Magellan_-_geometry_of_the_orbit.png** *Source:* https://upload.wikimedia.org/wikipedia/commons/4/40/Magellan_-_geometry_of_the_orbit.png *License:* Public domain *Contributors: Magellan: Mission to Venus,* JPL, 1991 *Original artist:* NASA / JPL

- **File:Magellan_-_imagery_for_VRM_and_from_past_missions.png** *Source:* https://upload.wikimedia.org/wikipedia/commons/4/4c/Magellan_-_imagery_for_VRM_and_from_past_missions.png *License:* Public domain *Contributors:* Dallas, *The Venus Radar Mapping Mission,* Acta Astronautica, JPL, p. 112 *Original artist:* NASA / JPL

- **File:Magellan_-_mapping_phase.png** *Source:* https://upload.wikimedia.org/wikipedia/commons/4/43/Magellan_-_mapping_phase.png *License:* CC BY-SA 3.0 *Contributors:* Own work *Original artist:* Xession

- **File:Magellan_-_radar_electronics.png** *Source:* https://upload.wikimedia.org/wikipedia/commons/6/68/Magellan_-_radar_electronics.png *License:* Public domain *Contributors:* The Magellan Venus explorer's guide[1], pg. 74 *Original artist:* NASA

- **File:Magellan_-_spacecraft_bus.png** *Source:* https://upload.wikimedia.org/wikipedia/commons/8/80/Magellan_-_spacecraft_bus.png *License:* Public domain *Contributors:* Magellan: Mission to Venus [1], pg. 11 *Original artist:* NASA / JPL

- **File:Magellan_-_trajectory.png** *Source:* https://upload.wikimedia.org/wikipedia/commons/9/9f/Magellan_-_trajectory.png *License:* CC BY-SA 3.0 *Contributors:* Own work *Original artist:* Xession

- **File:Magellan_Preparations.jpg** *Source:* https://upload.wikimedia.org/wikipedia/commons/6/65/Magellan_Preparations.jpg *License:* Public domain *Contributors:* Great Images in NASA Description *Original artist:* NASA

- **File:Magellan_Venus_globes.jpg** *Source:* https://upload.wikimedia.org/wikipedia/commons/a/a1/Magellan_Venus_globes.jpg *License:* Public domain *Contributors:* http://solarviews.com/cap/venus/venview.htm (image link) *Original artist:* NASA/JPL

- **File:Magellan_at_Kennedy_Space_Center.jpg** *Source:* https://upload.wikimedia.org/wikipedia/commons/1/1e/Magellan_at_Kennedy_Space_Center.jpg *License:* Public domain *Contributors:* ? *Original artist:* ?

- **File:Magellan_deploy.jpg** *Source:* https://upload.wikimedia.org/wikipedia/commons/6/62/Magellan_deploy.jpg *License:* Public domain *Contributors:* http://nssdc.gsfc.nasa.gov/photo_gallery/photogallery-spacecraft.html *Original artist:* NASA

- **File:Magellan_diagramm.png** *Source:* https://upload.wikimedia.org/wikipedia/commons/d/da/Magellan_diagramm.png *License:* Public domain *Contributors:* http://nssdc.gsfc.nasa.gov/nmc/spacecraftDisplay.do?id=1989-033B (http://nssdc.gsfc.nasa.gov/planetary/image/magellan_diagram.jpg) - Image converted to PNG and slightly enhanced to reduce grain *Original artist:* NASA

- **File:Magellan_orbit.jpg** *Source:* https://upload.wikimedia.org/wikipedia/commons/b/be/Magellan_orbit.jpg *License:* Public domain *Contributors:* [1]
Direct link *Original artist:* NASA / JPL

- **File:Magellan_to_venus.ogv** *Source:* https://upload.wikimedia.org/wikipedia/commons/b/b9/Magellan_to_venus.ogv *License:* Public domain *Contributors:* [1] *Original artist:* NASA / JPL

- **File:Mapa_de_sondas_sobre_Venus.png** *Source:* https://upload.wikimedia.org/wikipedia/commons/d/d9/Mapa_de_sondas_sobre_Venus.png *License:* Public domain *Contributors:* image:Map of Venus.png *Original artist:* User:KillOrDie

- **File:MarCO_CubeSat.jpg** *Source:* https://upload.wikimedia.org/wikipedia/commons/8/88/MarCO_CubeSat.jpg *License:* Public domain *Contributors:* http://www.airspacemag.com/daily-planet/cubesats-rescue-180955544/?no-ist *Original artist: This file is in the **public domain** in the United States because it was solely created by NASA. NASA copyright policy states that "NASA material is not protected by copyright **unless noted**". (See Template:PD-USGov, NASA copyright policy page or JPL Image Use Policy.)*

- **File:Mariner-10-Trajectory-first_half.PNG** *Source:* https://upload.wikimedia.org/wikipedia/commons/4/40/Mariner-10-Trajectory-first_half.PNG *License:* Public domain *Contributors: Mariner Venus-Mercury 1973 Project Final Report. Venus and Mercury I Encounters. Tecnical Memorandum 33-734 Volume I* (PDF - 18.1 MB) Pasadena, California, Jet Propulsion Laboratory, NASA, 9/15/1976, p. 2. *Original artist:* JPL, NASA

- **File:Mariner09.jpg** *Source:* https://upload.wikimedia.org/wikipedia/commons/1/11/Mariner09.jpg *License:* Public domain *Contributors:* http://nssdc.gsfc.nasa.gov/database/MasterCatalog?sc=1971-051A *Original artist:* NASA

- **File:Mariner10.jpg** *Source:* https://upload.wikimedia.org/wikipedia/commons/a/af/Mariner10.jpg *License:* Public domain *Contributors:*

- http://solarsystem.nasa.gov/multimedia/display.cfm?IM_ID=1563 *Original artist:* NASA / Jet Propulsion Laboratory

- **File:Mariner_10.jpg** *Source:* https://upload.wikimedia.org/wikipedia/commons/f/fe/Mariner_10.jpg *License:* Public domain *Contributors:* http://solarviews.com/cap/craft/marin10.htm (image link) *Original artist:* NASA

- **File:Mariner_10_1975_Issue-10c.jpg** *Source:* https://upload.wikimedia.org/wikipedia/commons/8/82/Mariner_10_1975_Issue-10c.jpg *License:* Public domain *Contributors:* US Post Office, Hi res scan of photo by Gwillhickers *Original artist:* US Post Office; Bureau of Engraving and Printing

- **File:Mariner_2.jpg** *Source:* https://upload.wikimedia.org/wikipedia/commons/9/90/Mariner_2.jpg *License:* Public domain *Contributors:* http://nix.larc.nasa.gov/info;jsessionid=22sidxj3uc1s0?id=PIA04594&orgid=10 *Original artist:* NASA Jet Propulsion Laboratory (NASA-JPL)

- **File:Mariner_2_launch.jpg** *Source:* https://upload.wikimedia.org/wikipedia/commons/0/04/Mariner_2_launch.jpg *License:* Public domain *Contributors:* http://history.nasa.gov/SP-480/ch8.htm *Original artist:* NASA

- **File:Mariner_3_and_4.jpg** *Source:* https://upload.wikimedia.org/wikipedia/commons/b/bc/Mariner_3_and_4.jpg *License:* Public domain *Contributors:*

- http://solarsystem.nasa.gov/multimedia/display.cfm?IM_ID=1897 *Original artist:* NASA

- **File:Mariner_5.jpg** *Source:* https://upload.wikimedia.org/wikipedia/commons/0/02/Mariner_5.jpg *License:* Public domain *Contributors:* Transfered from de.wikipedia Transfer was stated to be made by User:henristosch.
*Original artist:* NASA, Original uploader was W.wolny at de.wikipedia

- **File:Mariner_6and7.gif** *Source:* https://upload.wikimedia.org/wikipedia/commons/4/47/Mariner_6and7.gif *License:* Public domain *Contributors:* [1] *Original artist:* NASA

- **File:Mars-express-volcanoes-sm.jpg** *Source:* https://upload.wikimedia.org/wikipedia/commons/9/9d/Mars-express-volcanoes-sm.jpg *License:* Public domain *Contributors:* http://marsprogram.jpl.nasa.gov/express/gallery/artwork/marsis-radarpulses.html (image link) *Original artist:* NASA/JPL/Corby Waste

- **File:Pioneer_program.gif** *Source:* https://upload.wikimedia.org/wikipedia/commons/e/ec/Pioneer_program.gif *License:* Public domain *Contributors:* ? *Original artist:* ?

- **File:Portal-puzzle.svg** *Source:* https://upload.wikimedia.org/wikipedia/en/f/fd/Portal-puzzle.svg *License:* Public domain *Contributors:* ? *Original artist:* ?

- **File:Psyché20150929.jpg** *Source:* https://upload.wikimedia.org/wikipedia/commons/a/a1/Psych%C3%A920150929.jpg *License:* Public domain *Contributors:*

  http://www.jpl.nasa.gov/news/news.php?feature=4727

  *Original artist:* Image credit: NASA/JPL-Caltech"

- **File:Question_book-new.svg** *Source:* https://upload.wikimedia.org/wikipedia/en/9/99/Question_book-new.svg *License:* Cc-by-sa-3.0 *Contributors:*

  Created from scratch in Adobe Illustrator. Based on Image:Question book.png created by User:Equazcion *Original artist:*

  Tkgd2007

- **File:RocketSunIcon.svg** *Source:* https://upload.wikimedia.org/wikipedia/commons/d/d6/RocketSunIcon.svg *License:* Copyrighted free use *Contributors:* Self made, based on File:Spaceship and the Sun.jpg *Original artist:* Me

- **File:Roscosmos_logo_ru.svg** *Source:* https://upload.wikimedia.org/wikipedia/commons/d/da/Roscosmos_logo_ru.svg *License:* Public domain *Contributors:* Official site of the Russian Federal Space Agency *Original artist:* Russian Federal Space Agency

- **File:STS-30_launch.jpg** *Source:* https://upload.wikimedia.org/wikipedia/commons/5/56/STS-30_launch.jpg *License:* Public domain *Contributors:* NASA (deep-link ?) *Original artist:* NASA/exploitcorporations

- **File:Shuttle.svg** *Source:* https://upload.wikimedia.org/wikipedia/commons/a/a1/Shuttle.svg *License:* Public domain *Contributors:* ? *Original artist:* ?

- **File:Smallvenusprobe.gif** *Source:* https://upload.wikimedia.org/wikipedia/en/c/cc/Smallvenusprobe.gif *License:* Fair use *Contributors:*

  Annual Big Soviet Encyclopaedia Journal

  *Original artist:* ?

- **File:Symbol_book_class2.svg** *Source:* https://upload.wikimedia.org/wikipedia/commons/8/89/Symbol_book_class2.svg *License:* CC BY-SA 2.5 *Contributors:* Mad by Lokal_Profil by combining: *Original artist:* Lokal_Profil

- **File:Text_document_with_red_question_mark.svg** *Source:* https://upload.wikimedia.org/wikipedia/commons/a/a4/Text_document_with_red_question_mark.svg *License:* Public domain *Contributors:* Created by bdesham with Inkscape; based upon Text-x-generic.svg from the Tango project. *Original artist:* Benjamin D. Esham (bdesham)

- **File:Tmk-mavr.jpg** *Source:* https://upload.wikimedia.org/wikipedia/commons/8/8c/Tmk-mavr.jpg *License:* Public domain *Contributors:* Para el diseño de la nave, he tomado como referencia las ilustraciones que se pueden ver en http://www.friends-partners.org/partners/mwade/craft/mavr.htm aunque el diseño no me ha salido al 100% igual, pero yo creo más o menos *"se le parece"* . La imagen de Venus que se ve de fondo es la que hay en http://commons.wikimedia.org/wiki/Image:Venus-real_color.jpg *Original artist:* User:KillOrDie

- **File:Translation_to_english_arrow.svg** *Source:* https://upload.wikimedia.org/wikipedia/commons/8/8a/Translation_to_english_arrow.svg *License:* CC-BY-SA-3.0 *Contributors:* Transferred from en.wikipedia; transferred to Commons by User:Faigl.ladislav using CommonsHelper. *Original artist:* tkgd2007. Original uploader was Tkgd2007 at en.wikipedia

- **File:US-Satellite.svg** *Source:* https://upload.wikimedia.org/wikipedia/commons/2/2f/US-Satellite.svg *License:* Public domain *Contributors:* Created myself using Inkscape, incorporates PD File:Earth clip art.svg and File:Flag of the United States.svg *Original artist:* **<a href='//en.wikipedia.org/wiki/User:GW_Simulations' class='extiw' title='en:User:GW Simulations'>G</a><a href='//commons.wikimedia.org/wiki/User_talk:GW_Simulations' title='User talk:GW Simulations' class='mw-redirect'>W</a>** ··· (User • Talk • EN)

- **File:USSR-Satellite.svg** *Source:* https://upload.wikimedia.org/wikipedia/commons/4/4f/USSR-Satellite.svg *License:* Public domain *Contributors:* Created myself using Inkscape, incorporates PD File:Earth clip art.svg and File:Flag of the Soviet Union.svg *Original artist:* **<a href='//en.wikipedia.org/wiki/User:GW_Simulations' class='extiw' title='en:User:GW Simulations'>G</a><a href='//commons.wikimedia.org/wiki/User_talk:GW_Simulations' title='User talk:GW Simulations' class='mw-redirect'>W</a>** ··· (User • Talk • EN)

- **File:Vega-mission.jpg** *Source:* https://upload.wikimedia.org/wikipedia/commons/1/13/Vega-mission.jpg *License:* Public domain *Contributors:* ? *Original artist:* ?

- **File:Vega_model_-_Udvar-Hazy_Center.JPG** *Source:* https://upload.wikimedia.org/wikipedia/commons/5/52/Vega_model_-_Udvar-Hazy_Center.JPG *License:* Public domain *Contributors:* Own work *Original artist:* Daderot

- **File:Venera9.png** *Source:* https://upload.wikimedia.org/wikipedia/en/1/15/Venera9.png *License:* ? *Contributors:*

  http://www.mentallandscape.com/C_CatalogVenus.htm *Original artist:*

  Soviet Union/Roscosmos/Venera 9

- **File:Venera_10_orbiter.jpg** *Source:* https://upload.wikimedia.org/wikipedia/commons/6/6f/Venera_10_orbiter.jpg *License:* Public domain *Contributors:* ? *Original artist:* ?

- **File:Venera_13_lander.gif** *Source:* https://upload.wikimedia.org/wikipedia/commons/c/c7/Venera_13_lander.gif *License:* Public domain *Contributors:* ? *Original artist:* ?

- **File:Venera_13_orbiter.gif** *Source:* https://upload.wikimedia.org/wikipedia/commons/6/61/Venera_13_orbiter.gif *License:* Public domain *Contributors:* ? *Original artist:* ?

- **File:Venera_14_-_venera14.jpg** *Source:* https://upload.wikimedia.org/wikipedia/commons/9/95/Venera_14_-_venera14.jpg *License:* Public domain *Contributors:* Gallery Link, Photo Link *Original artist:* USSR / Preserved by the NASA National Space Science Data Center

- **File:Venera_15.gif** *Source:* https://upload.wikimedia.org/wikipedia/commons/a/a6/Venera_15.gif *License:* Public domain *Contributors:* http://nssdc.gsfc.nasa.gov/database/MasterCatalog?sc=1983-053A *Original artist:* NASA

- **File:Venera_1962_diagramm.jpg** *Source:* https://upload.wikimedia.org/wikipedia/commons/1/1d/Venera_1962_diagramm.jpg *License:* Public domain *Contributors:* ? *Original artist:* ?

- **File:Venera_4.jpg** *Source:* https://upload.wikimedia.org/wikipedia/commons/9/9c/Venera_4.jpg *License:* Public domain *Contributors:* ? *Original artist:* ?

- **File:Venera_7.jpg** *Source:* https://upload.wikimedia.org/wikipedia/commons/b/b4/Venera_7.jpg *License:* Public domain *Contributors:* Own work *Original artist:* Rafael323

- **File:Venera_8.jpg** *Source:* https://upload.wikimedia.org/wikipedia/commons/b/b7/Venera_8.jpg *License:* Public domain *Contributors:* ? *Original artist:* ?

- **File:Venera_8_capsule.jpg** *Source:* https://upload.wikimedia.org/wikipedia/commons/6/6e/Venera_8_capsule.jpg *License:* Public domain *Contributors:* ? *Original artist:* ?

- **File:Venera_9_-_Venera_10_-_venera9-10.jpg** *Source:* https://upload.wikimedia.org/wikipedia/commons/e/ee/Venera_9_-_Venera_10_-_venera9-10.jpg *License:* Public domain *Contributors:* Gallery Link, Photo Link, [1] *Original artist:* USSR / Preserved by the NASA National Space Science Data Center

- **File:Venera_9_lander.jpg** *Source:* https://upload.wikimedia.org/wikipedia/commons/a/a1/Venera_9_lander.jpg *License:* Public domain *Contributors:* ? *Original artist:* ?

- **File:Venera_9_orbiter.jpg** *Source:* https://upload.wikimedia.org/wikipedia/commons/d/df/Venera_9_orbiter.jpg *License:* Public domain *Contributors:* ? *Original artist:* ?

- **File:Venus,_Earth_size_comparison.jpg** *Source:* https://upload.wikimedia.org/wikipedia/commons/9/92/Venus%2C_Earth_size_comparison.jpg *License:* Public domain *Contributors:* The Earth seen from Apollo 17.jpg
  Venus globe.jpg *Original artist:* NASA
  Venus image: NASA / JPL (from *Magellan*)

- **File:Venus-real.jpg** *Source:* https://upload.wikimedia.org/wikipedia/commons/5/51/Venus-real.jpg *License:* Copyrighted free use *Contributors:* http://astrosurf.com/nunes/explor/explor_m10.htm *Original artist:* NASA/Ricardo Nunes

- **File:VenusAnimation.ogg** *Source:* https://upload.wikimedia.org/wikipedia/commons/0/06/VenusAnimation.ogg *License:* Public domain *Contributors:* Taken from the Magellan Probe. http://laps.noaa.gov/albers/sos/venus/venus4/venus4_rgb_cyl_www.jpg *Original artist:* Rendered by Ironchew, image courtesy of NASA.

- **File:VenusFlybyCutaway.jpg** *Source:* https://upload.wikimedia.org/wikipedia/en/4/49/VenusFlybyCutaway.jpg *License:* Fair use *Contributors:*
  http://ntrs.nasa.gov/archive/nasa/casi.ntrs.nasa.gov/19790072165_1979072165.pdf *Original artist:*
  Bellcomm, Inc

- **File:VenusFlybyStudyMissionPhases.jpg** *Source:* https://upload.wikimedia.org/wikipedia/en/8/84/VenusFlybyStudyMissionPhases.jpg *License:* Fair use *Contributors:*
  http://ntrs.nasa.gov/archive/nasa/casi.ntrs.nasa.gov/19790072165_1979072165.pdf *Original artist:*
  Bellcomm, Inc

- **File:VenusLanderTopo.jpg** *Source:* https://upload.wikimedia.org/wikipedia/commons/9/98/VenusLanderTopo.jpg *License:* CC BY-SA 3.0 *Contributors:*

- Pioneer *Original artist:* Zamonin

- **File:VenusTopoVenera.jpg** *Source:* https://upload.wikimedia.org/wikipedia/commons/8/8e/VenusTopoVenera.jpg *License:* CC BY-SA 3.0 *Contributors:* Own work *Original artist:* Zamonin

- **File:Venus_-_3D_Perspective_View_of_Maat_Mons.jpg** *Source:* https://upload.wikimedia.org/wikipedia/commons/8/8c/Venus_-_3D_Perspective_View_of_Maat_Mons.jpg *License:* Public domain *Contributors:* http://photojournal.jpl.nasa.gov/catalog/PIA00106 *Original artist:* This image or video was catalogued by Jet Propulsion Laboratory of the United States National Aeronautics and Space Administration (NASA) under **Photo ID:** PIA00106.

- **File:Venus_Express_in_orbit_(crop).jpg** *Source:* https://upload.wikimedia.org/wikipedia/commons/0/0b/Venus_Express_in_orbit_%28crop%29.jpg *License:* Public domain *Contributors:* Self-made with Celestia program with Addon by Jestr *Original artist:* Andrzej Mirecki

- **File:Venus_In-Situ_Explorer.png** *Source:* https://upload.wikimedia.org/wikipedia/commons/5/54/Venus_In-Situ_Explorer.png *License:* Public domain *Contributors:* http://newfrontiers.nasa.gov/images/2_06_VENUS_lg.png *Original artist:* NASA

- **File:Venus_clouds_seen_by_Pioneer_Venus_Orbiter.png** *Source:* https://upload.wikimedia.org/wikipedia/commons/d/d4/Venus_clouds_seen_by_Pioneer_Venus_Orbiter.png *License:* Public domain *Contributors:* http://nssdc.gsfc.nasa.gov/photo_gallery/photogallery-venus.html direct link to the picture:ftp://nssdcftp.gsfc.nasa.gov/photo_gallery/hi-res/planetary/venus/pvo_uv_790205.tiff *Original artist:* Pioneer Venus Orbiter (NASA)

- **File:Venus_dome_3D.jpg** *Source:* https://upload.wikimedia.org/wikipedia/commons/4/44/Venus_dome_3D.jpg *License:* Public domain *Contributors:* ? *Original artist:* ?

- **File:Venus_globe.jpg** *Source:* https://upload.wikimedia.org/wikipedia/commons/8/85/Venus_globe.jpg *License:* Public domain *Contributors:* http://photojournal.jpl.nasa.gov/catalog/PIA00104 *Original artist:* NASA

- **File:Venus_map_with_labels.jpg** *Source:* https://upload.wikimedia.org/wikipedia/commons/d/dd/Venus_map_with_labels.jpg *License:* Public domain *Contributors:* http://ails.arc.nasa.gov/ails/?v=thumbs&st=1&so=unsorted&page=1&r=0&qs=AC78-9135&x=0&y=0 *Original artist:* NASA Ames Reseach Center, U.S Geological Survey and Messachusetts Institute of Technology

- **File:Venuspioneeruv.jpg** *Source:* https://upload.wikimedia.org/wikipedia/commons/b/bc/Venuspioneeruv.jpg *License:* Public domain *Contributors:* NSSDC Photo Gallery Venus direct link to the big TIFF Version:ftp://nssdcftp.gsfc.nasa.gov/photo_gallery/hi-res/planetary/venus/pvo_uv_790226.tiff *Original artist:* NASA

- **File:Venusvulkan_Tick-Typ.jpg** *Source:* https://upload.wikimedia.org/wikipedia/commons/d/d7/Venusvulkan_Tick-Typ.jpg *License:* Public domain *Contributors:*

- NASA planetary photojournal, PIA00089 *Original artist:* NASA/JPL

- **File:Veritas20150930.jpg** *Source:* https://upload.wikimedia.org/wikipedia/commons/c/c9/Veritas20150930.jpg *License:* Public domain *Contributors:* http://www.jpl.nasa.gov/news/news.php?feature=4727

  *Original artist:* Image credit: NASA/JPL-Caltech

- **File:Video-mariner-2-launch-020429.ogg** *Source:* https://upload.wikimedia.org/wikipedia/commons/5/5b/Video-mariner-2-launch-020429.ogg *License:* Public domain *Contributors:* http://www.jpl.nasa.gov/history/60s/Mariner1962.html *Original artist:* NASA

- **File:Vénus_télescope.jpg** *Source:* https://upload.wikimedia.org/wikipedia/commons/c/cc/V%C3%A9nus_t%C3%A9lescope.jpg *License:* GFDL *Contributors:* http://astrosurf.com/lecleire/2007/venuscolor_100907_05h22.jpg *Original artist:* Marc Lecleire

- **File:WT-reentry-fireball-tight.jpg** *Source:* https://upload.wikimedia.org/wikipedia/commons/c/c6/WT-reentry-fireball-tight.jpg *License:* Public domain *Contributors:* WT1190F Airborne observation team *Original artist:* NASA/IAC/UAE Space Agency

- **File:Wikinews-logo.svg** *Source:* https://upload.wikimedia.org/wikipedia/commons/2/24/Wikinews-logo.svg *License:* CC BY-SA 3.0 *Contributors:* This is a cropped version of Image:Wikinews-logo-en.png. *Original artist:* Vectorized by Simon 01:05, 2 August 2006 (UTC) Updated by Time3000 17 April 2007 to use official Wikinews colours and appear correctly on dark backgrounds. Originally uploaded by Simon.

- **File:Wild2_3.jpg** *Source:* https://upload.wikimedia.org/wikipedia/commons/4/41/Wild2_3.jpg *License:* Public domain *Contributors:*

- http://nssdc.gsfc.nasa.gov/database/MasterCatalog?sc=1999-003A *Original artist:* NASA

- **File:Zond1.gif** *Source:* https://upload.wikimedia.org/wikipedia/commons/0/07/Zond1.gif *License:* Public domain *Contributors:* [1] *Original artist:* NASA

- **File:Zond_2.jpg** *Source:* https://upload.wikimedia.org/wikipedia/commons/e/ee/Zond_2.jpg *License:* No restrictions *Contributors:* Zond 2 *Original artist:* NASA on The Commons

- **File:Zond_L1_drawing.png** *Source:* https://upload.wikimedia.org/wikipedia/commons/1/15/Zond_L1_drawing.png *License:* Public domain *Contributors:* http://ston.jsc.nasa.gov/collections/TRS/_techrep/RP1357.pdf *Original artist:* NASA

## 2.3 Content license